Space in Weak Propositional Proof Systems

Ilario Bonacina

Space in Weak Propositional Proof Systems

 Springer

Ilario Bonacina
Departament de Ciències de la Computació
Universitat Politècnica de Catalunya
Barcelona
Spain

ISBN 978-3-319-89249-8   ISBN 978-3-319-73453-8   (eBook)
https://doi.org/10.1007/978-3-319-73453-8

Printed on acid-free paper

This Springer imprint is published by the registered company Springer International Publishing AG part
of Springer Nature
The registered company address is: Gewerbestrasse 11, 6330 Cham, Switzerland

to my family

Preface

In this book you, the reader, are going to see some results on the *space* complexity of some propositional proof systems. This book is a revised version of my PhD thesis[1] and indeed it is not intended to be a survey of *all* the results known on the space complexity of propositional proof systems. It will rather be a long walk touching some topics in proof complexity, mostly about space of course, but not exclusively. Hopefully this could be used too as a rather reader-friendly exposition of some game theoretic methods used in proof complexity. This is indeed an underground theme connecting most of the results we show. Of course there will be some survey(-ish) parts but mainly the focus will be on the new game theoretic techniques and their application to the analysis of the space complexity of propositional proof systems. That is the results arising from my PhD thesis [Bon15] and some earlier works [BG13, BGT14, BT15, BG15, BBG+17, BT16a, BT16b, BGT16, Bon16].

This is a work about proof complexity, so let's start by introducing it informally. *Proof complexity* is a research area that studies the concept of complexity from the point of view of logic. In particular, in proof complexity we are interested in questions such as: *"how difficult is it to prove a theorem?"* Or, more precisely, given a formal system, we are interested in measuring the complexity of a theorem, that is answering questions such as *"what is the shortest proof of the theorem in a given formal system?"* This mirrors questions in computational complexity about, for example, the number of steps that a Turing machine needs to compute a given function f; or the size of circuits needed to compute f.[2]

In this book we investigate the space complexity of propositional proof systems, so what is the *space* of a proof? We could state this question pictorially as *"what is the smallest blackboard a teacher needs to present the proof of a theorem to a class of students?"*[3] As before, this notion is analogous to the space complexity

---

[1] This revised version is due to the fact that my thesis was awarded *"Best Italian PhD Thesis in Theoretical Computer Science"* for the year 2016 by the Italian chapter of the European Association for Theoretical Computer Science (EATCS).

[2] On the other hand, we could also measure the complexity of a theorem as the strength of a theory needed to prove the theorem. This also has a counterpart in computational complexity, it is linked with questions about the smallest complexity class to which a given function belongs.

[3] We suppose here that the students can understand just proofs written on the blackboard in some given formal system and they do not have any additional memory except the minimal one to

in the context of uniform computations, measured, for example, as the size of a working-tape needed by a Turing machine to compute a given function.

Propositional proof complexity, that is the complexity of propositional proofs, plays a role in the context of feasible proofs as important as the role of Boolean circuits in the context of efficient computations. Although the original motivations to study the complexity of propositional proofs came from proof-theoretical questions about first-order theories, it turns out that, essentially, the complexity of propositional proofs deals with the following question: *"what can be proved by a prover with bounded computational abilities?"* For example, if its computational abilities are limited to small circuits from some circuit class. Hence, propositional proof complexity mirrors non-uniform computational complexity and indeed there is a very productive cross-fertilization of techniques between the two fields. Our understanding of propositional proof systems is, unfortunately, similar to the general situation in complexity theory. In both fields we can prove lower bounds in very special cases and indeed there are several major open problems that are very basic, way more basic than the well-known question $P \overset{?}{=} NP$. The situation is similar in the sense that we can prove super-polynomial lower bounds on the length of proofs only for restricted proof systems. Indeed, by a result of [CR79], proving super-polynomial lower bounds on the length of proofs for *every* propositional proof system is equivalent to showing that $NP \neq coNP$, which in turn is one of the open and very important problems in computational complexity. Propositional proof complexity is important also from the practical point of view. The implementations of state-of-the-art SAT algorithms ultimately rely on rather simple propositional proof systems. Hence the study of those systems helps in clarifying the limitations of such algorithms that are essential in various aspects of computer science, cf. [Nor15].

We will focus on the space complexity of two particular proof systems: *resolution*, a well studied proof system that is at the core of state-of-the-art SAT-solvers; and *polynomial calculus*, a proof system that uses polynomials to refute propositional formulas that are contradictions. We will show some generic combinatorial techniques to prove space lower bounds in both those systems and then we will apply those techniques to show concrete space lower bounds for refutations of several particular (unsatisfiable) propositional formulas. Since the very first exponential size lower bound for resolution size in [Hak85], game theoretic methods and combinatorial characterisations of hardness measures have a long history in proof complexity. This book could be seen as the latest contribution to this topic.

For resolution the new techniques we introduce allowed for the first time to obtain—in a quite easy way actually—lower bounds for the space of proofs when the space is measured as the total number of variables to be kept in memory (*total space*). For polynomial calculus the techniques we introduce—which is more involved than those for resolution—allow us to address space lower bounds when the space takes into account the number of distinct monomials to be kept in memory (*monomial space*). Notably those techniques allow us to prove, among other results, that almost

---

understand the content of the blackboard. Moreover the teacher has to write with fonts of a fixed size.

all k-CNF formulas are hard with respect to total space in resolution and monomial space in polynomial calculus. That is we prove asymptotically optimal lower bounds for the monomial space (and total space in resolution) for random k-CNF formulas in n variables and a linear number of clauses. This was an open problem mentioned for the first time in [BS01, ABRW02] and since then reported many times in the literature.

Book Structure After an introduction to propositional proof complexity (Chap. 1), this work consists of 3 parts. Each chapter ends with a section containing open questions and a **History** section collecting some facts about the main theorems of the chapter and how they fit in the previous literature.

In **Part I** there are two chapters on resolution: one containing results already known in the literature before this work (Chap. 2) and one just focused on space in resolution (Chap. 3). More precisely on the combinatorial techniques to prove total space lower bounds. Then we move to polynomial calculus and its space complexity (Chap. 4). The focus will be now on the combinatorial technique to prove monomial space lower bounds.

In **Part II** we collect the main applications of the techniques we built previously. First there is a short chapter about the proof complexity and space complexity of the pigeon principles (PHP_n^m and its variations), cf. Chap. 5. Then there is an interlude on some new type of games, the *cover games*, defined on bipartite graphs (Chap. 6). This chapter is essentially independent from the rest of the book and it collects some results on graph theory. The motivation behind this chapter though is that the results in it will be needed in Chap. 7 to prove the space lower bound for random k-CNF formulas and other graph-based propositional formulas.

In the last part, **Part III**, we analyse the size of resolution proofs in connection with the *Strong Exponential Time Hypothesis* (SETH) in complexity theory. More precisely we prove strong size lower bounds for a restricted version of resolution we call δ-regular resolution. Although not directly related to space, the results we show here rely on some combinatorial characterisations and games analogous to the one used to prove space lower bounds.

Acknowledgments First of all, I want to thank my advisor Nicola Galesi at Sapienza University of Rome. He cultivated my interest in proof complexity supporting the good ideas and shooting down the (many) buggy ones. This book simply wouldn't have existed without him.

I am truly grateful to the many people from the proof complexity community that I had the privilege to meet. My view of proof complexity and theoretical computer science was shaped by all of the discussions we had. In particular I want to thank (the order is not relevant here): Jakob Nordström, Massimo Lauria, Mladen Mikša, Marc Vinyals, Susanna Figueiredo de Rezende, Navid Talebanfard, Tony Huynh, Paul Wollan, Pavel Pudlák, Jan Krajíček, Neil Thapen, Olaf Beyersdorff, Rahul Santhanam, Albert Atserias, Jacobo Torán, and Yuval Filmus.

Stockholm,   *Ilario Bonacina*
May 31, 2017

Contents

List of Figures

Notation

In this section we give the notation that shall be standard throughout this book.

Sets

We use the standard set-theoretic notations.

- $|S|$ is the size of the set S
- $[n]$ is the set of natural numbers $\{1,\dots,n\}$
- $\emptyset$ is the empty set
- $A \cup B = \{x \ : \ x \in A \text{ or } x \in B\}$
- $A \cap B = \{x \ : \ x \in A \text{ and } x \in B\}$
- $A \dot\cup B = A \cup B$ in the case $A \cap B = \emptyset$
- $A \setminus B = \{x \ : \ x \in A \text{ and } x \notin B\}$
- $A \subseteq B$ if every element of A is also an element of B
- (a,b) is an ordered pair of elements
- $A \times B = \{(x,y) \ : \ x \in A \text{ and } y \in B\}$
- $\binom{S}{2}$ is the set of subsets of the set S of size 2

Arithmetic

As customary, $\mathbb{N}$ is the set of all natural numbers, $\mathbb{R}$ is the set of real numbers, $\mathbb{F}$ is a generic field and $\mathbb{F}_p$ is a finite Galois field with p elements. Given a field $\mathbb{F}$, $\mathrm{char}(\mathbb{F})$ is the smallest integer a such that for every element x of $\mathbb{F}$, $\underbrace{x + \cdots + x}_{a} = 0$.

If not stated otherwise e will be the base of natural logarithms, $e = 2.718\dots$ We denote as $\ln(\cdot)$ the natural logarithm and with $\log(\cdot)$ the logarithm over base 2. Given a real number x, $\lfloor x \rfloor$ is the largest integer smaller or equal to x. The binomial coefficient $\binom{m}{n}$ is $\frac{m!}{n!(m-n)!}$. We use sometimes the inequality $\binom{m}{n} \leqslant \left(\frac{em}{n}\right)^n$.

Asymptotic notations. Given two functions f, g from $\mathbb{N}$ to $\mathbb{N}$ we say that $f = O(g)$ if there are some absolute constants c_1, c_2 such that for every $n \in \mathbb{N}$, $f(n) \leqslant c_1 g(n) + c_2$. We say that $f = \Omega(g)$ if $g = O(f)$ and $f = \Theta(g)$ if both $f = \Omega(g)$ and $f = O(g)$. We say that $f = \widetilde{O}(g)$ if there exists a $k \in \mathbb{N}$ such that $f = O\left(g \log^k g\right)$. We say that $f = o(g)$ if $\frac{f(n)}{g(n)} \to 0$ as $n \to \infty$. We say that $f = \omega(g)$ if $g = o(f)$. We say that f is *super-polynomial* in n if $f = n^{\omega(1)}$.

Logic

Propositional formulas. A Boolean variable x and its negation $\neg x$ are sometimes denoted respectively as x^1 and x^0. A *literal* ℓ is a Boolean variable or the negation of a Boolean variable. A disjunction of literals $\bigvee_{i \in I} \ell_i$ is a *clause*. Its size $|C|$ is the number of distinct literals in C. If $|C| \leqslant k$ we say that C is a *k-clause*. A conjunction of clauses $\{C_i : i \in [m]\}$ is a formula in *Conjunctive Normal Form* (CNF formula) and it is denoted also as $C_1 \wedge \cdots \wedge C_m$. If all the clauses are k-clauses then we say that the formula is a *k-CNF formula*. Given a CNF formula F, the set of Boolean variables mentioned in F is $\mathrm{vars}(F)$. The number of clauses mentioned in the CNF formula F is $|F|$.

We often consider families of formulas $(F_n)_{n \in \mathbb{N}}$ where usually $n = |\mathrm{vars}(F_n)|$ or n is polynomially related to $|\mathrm{vars}(F_n)|$. With a slight abuse of notation a family of formulas $(F_n)_{n \in \mathbb{N}}$ is denoted simply as F_n.

Boolean assignments. Given a set of variables X, a *Boolean assignment* over X is a map $\alpha : X \to \{0, 1, \star\}$, where X is a set of variables. The *domain* of α is $\mathrm{dom}(\alpha) = \alpha^{-1}(\{0, 1\})$. We say that α is *assigning* a value to x if and only if $x \in \mathrm{dom}(\alpha)$. With λ we denote the unique Boolean assignment with empty domain.

Given a Boolean assignments α over X and α' over X', their *union* $\alpha \cup \alpha'$ is the following Boolean assignment over $X \cup X'$

$$\alpha \cup \alpha'(x) = \begin{cases} \alpha(x) & \text{if } x \in \mathrm{dom}(\alpha) \setminus \mathrm{dom}(\alpha') \\ \alpha'(x) & \text{if } x \in \mathrm{dom}(\alpha') \setminus \mathrm{dom}(\alpha) \\ \alpha(x) & \text{if } x \in \mathrm{dom}(\alpha) \cap \mathrm{dom}(\alpha') \text{ and } \alpha(x) = \alpha'(x) \\ \star & \text{otherwise .} \end{cases} \tag{0.1}$$

Given a Boolean assignment α over X and $Y \subseteq X$, the *restriction* $\alpha\!\restriction_Y$ is the Boolean assignment

$$\alpha\!\restriction_Y(x) = \begin{cases} \alpha(x) & \text{if } x \in Y \\ \star & \text{otherwise .} \end{cases} \tag{0.2}$$

Given two Boolean assignments α and α', we say that $\alpha \subseteq \alpha'$, if $\alpha'\!\restriction_{\mathrm{dom}(\alpha)} = \alpha$.

Evaluation of CNF formulas. Given a CNF formula F and a Boolean assignment α over $\mathrm{vars}(F)$, we can apply α to F obtaining a new CNF formula $F\!\restriction_\alpha$ in this way: for each variable $x \in \mathrm{dom}(\alpha)$ substitute x in F with the value $\alpha(x)$, otherwise leave

x untouched. Then simplify the resulting formula with the usual rules: $0 \vee C \equiv C$, $1 \vee C \equiv 1$, $0 \wedge C \equiv 0$, $1 \wedge C \equiv C$. We say that α *satisfies* F if $F{\restriction}_\alpha = 1$. We denote this as $\alpha \vDash F$. Similarly, for a family A of Boolean assignments we write $A \vDash F$ if for each $\alpha \in A$, $\alpha \vDash F$.

Algebra

Given a field $\mathbb{F}$ and a set of variables X, the ring $\mathbb{F}[X]$ is the ring of polynomials in the variables X with coefficients in $\mathbb{F}$. An *ideal* I in $\mathbb{F}[X]$ is any subset of $\mathbb{F}[X]$ closed under addition, $p, q \in I$ implies that $p + q \in I$, and closed under multiplication with elements of $\mathbb{F}[X]$, $p \in I$ and $q \in \mathbb{F}[X]$ implies that $pq \in I$. Given a set of polynomials P, $\langle P \rangle$ is the ideal generated by P in $\mathbb{F}[X]$. Given two ideals I, J in $\mathbb{F}[X]$, $I + J = \{a + b \ : \ a \in I$ and $b \in J\}$.

Evaluations of polynomials. Given a polynomial p in $\mathbb{F}[X]$ and a Boolean assignment α we define the *restriction* $p{\restriction}_\alpha$, as follows: for each variable $x \in \mathrm{dom}(\alpha)$ substitute x in p with the value $\alpha(x)$, or otherwise leave the variable untouched. Then simplify the result with the usual simplification rules including: $0 \cdot m \equiv 0$, $1 \cdot m \equiv m$ and $m - m \equiv 0$ where m is any term in p, that is any monomial with a coefficient from $\mathbb{F}$ in front of it.

Graphs

A *graph* G is a pair (V, E) where V is a set and $E \subseteq \binom{V}{2}$. The elements of V are called *vertices* of G and the elements of E are called *edges* of G. Given a vertex $v \in G$, the *neighbor* of v in G is $N_G(v) = \{w \in V \ : \ \{v, w\} \in E\}$. The size of $N_G(v)$ is the *degree* of v in G.

A graph G is a *bipartite graph* if there exists two disjoint sets L and U such that $V = L \dot{\cup} U$ and $E \subseteq \{\{v, w\} \ : \ v \in L$ and $w \in U\}$. The pair (L, U) is a *bipartition* of the bipartite graph G.

A *matching* in G is a set $M \subseteq E$ such that all the edges in M are pair-wise disjoint. A matching *covers* $S \subseteq V$ if for each $v \in S$ there exists $e \in M$ such that $v \in e$.

A standard result about matchings in bipartite graphs is **Hall's theorem**: given any bipartite graph G with bipartition (L, U), the following are equivalent

1. G has a matching covering L;
2. for every subset $S \subseteq L$, $|N_G(S)| \geqslant |S|$.

Bipartite expansion. Let $r \in \mathbb{N}$ and $c \in \mathbb{R}$. A bipartite graph G with bipartition (L, U) is a (r, c)-*bipartite expander* if and only if

$$\forall A \subseteq L(G), |A| \leqslant r \rightarrow |N_G(A)| \geqslant c|A| \ . \tag{0.3}$$

Chapter 1
Introduction

This chapter is a general introduction to some of the themes of propositional proof complexity. We introduce on a high level proof systems such as resolution and polynomial calculus, we recall some connections between proof complexity and SAT algorithms, and we introduce on a high level the topic of space in propositional proof systems. We conclude with a summary of the results shown in this book.

1.1 Propositional Proof Systems

Formally, we consider *proofs* and *theorems* as strings over some finite alphabet, say strings in $\{0,1\}^*$. Then, following [CR79], we can define a proof system as follows.

Definition 1.1 (Proof System). A *proof system* for a language $\mathcal{L} \subseteq \{0,1\}^*$ is a polynomial-time onto function $P : \{0,1\}^* \to \mathcal{L}$.

Each string $T \in \mathcal{L}$ is a *theorem* and if $P(\pi) = T$, π is a *proof* of T in P, or a P-*proof* of T. Given a polynomial-time function $P : \{0,1\}^* \to \{0,1\}^*$ the fact that $P(\{0,1\}^*) \subseteq \mathcal{L}$ is the *soundness* property of P and the fact that $P(\{0,1\}^*) \supseteq \mathcal{L}$ is the *completeness* property of P.

The proof systems we consider are *propositional*, that is they are proof systems for the language UNSAT of unsatisfiable propositional formulas over Boolean variables that are in Conjunctive Normal Form (CNF).[1] That is Boolean formulas that are a conjunction ($\wedge$) of disjunctions ($\vee$), of variables and negated variables.

The computational complexity of a proof system for a language $\mathcal{L}$ varies a lot depending on the language $\mathcal{L}$ itself. It can vary from easy, say P, for a language $\mathcal{L}$ in NP; to coNP for the language UNSAT; to PSPACE for the language of True (fully) Quantified Boolean Formulas (TQBF); or it can be completely intractable, for example when $\mathcal{L}$ is First-Order Logic (FO), due to the recursive undecidability of the existence of proofs in FO.

---

[1] Equivalently, propositional proof systems could be defined for the coNP complete language TAUT of tautologically true propositional Boolean formulas.

© Springer International Publishing AG, part of Springer Nature 2017
I. Bonacina, *Space in Weak Propositional Proof Systems*,
https://doi.org/10.1007/978-3-319-73453-8_1

In this book we focus on particular examples of propositional proof systems and it will be helpful sometimes to compare their strength. This is done through the notion of p-simulation [CR79].

Definition 1.2 (p-Simulation). Given two propositional proof systems P and Q we say that P *p-simulates* Q if there exists a polynomial-time function t such that for each $\pi \in \{0,1\}^*$, $P(t(\pi)) = Q(\pi)$. Two systems are called *p-equivalent* if they p-simulate each other. If Q p-simulates P and there exist some formulas requiring exponentially long proofs in P but polynomially long proofs in Q we say that P *is exponentially weaker than Q*.

The main open problem in propositional proof complexity is about the length of proofs and, in particular, it concerns proving (or more likely disproving) the existence of a propositional proof system where all proofs are *polynomially bounded*.

Theorem 1.1 ([CR79]). *The following are equivalent:*

1. $NP = coNP$; *and*
2. *there exists a propositional proof system P that is* p-bounded, *i.e., such that there exists a polynomial p and every $F \in UNSAT$ has a P-proof of length at most $p(|F|)$, where $|F|$ is the length of F.*   □

Since it is usually conjectured that $NP \neq coNP$, the main goal of propositional proof complexity is actually to show that p-bounded propositional proof systems actually do not exist. One approach to this problem is to show that particular proof systems of increasing strength are not p-bounded. This approach is sometimes called *Cook's program* in proof complexity (although apparently Stephen Cook never proposed it explicitly). Indeed, this approach is somewhat unusual but it has some appeal in the following sense:

> "*Proving that* NP $\neq$ coNP *showing incrementally that examples of proof systems are not polynomially bounded seems unlikely. Rarely a universal statement is proved by proving all its instances. Nevertheless proving these lower bounds we may hope to uncover hidden computational hardness assumptions and then try to reduce the conjecture to some more approachable problem.*" [Kra09]

Now, how do we show that a particular propositional proof system P is not p-bounded? To do so it is sufficient to provide, even non-constructively, some family of formulas $(F_n)_{n\in\mathbb{N}}$ such that the minimal length of a proof of F_n in P grows super-polynomially with respect to $|F_n|$. Usually such formulas are propositional encodings of quite easy well-understood combinatorial principles, for example the pigeonhole principle PHP_n^m, see Sect. 5.1. We will see in detail some other examples of such Boolean formulas in Part II of this book but for the moment let's continue the general overview of propositional proof systems.

As we already said, we will focus on two particular propositional proof systems. Those are *resolution*, a logic one, and *polynomial calculus*, an algebraic one. But before doing this we want to give an idea of the richness of the landscape of proof systems studied in propositional proof complexity. There are the ones the reader might expect, the logic-based ones that are the common 'textbook' proof systems (e.g.

Frege systems), then there are, for example, some based on algebraic reasoning (e.g., *Polynomial Calculus* or *Sum-of-Squares*), geometric reasoning (e.g., *Cutting Planes* or the *Lovász-Schrijver* calculus) or based on some graph theoretic constructions (e.g., *Hajoś Calculus* [PU95]). We will overview briefly some of the propositional proof systems based on logic inference rules, or algebraic or geometric inference rules.

1.1.1 Frege Systems

Frege systems are the first propositional proof systems that *everybody* encounters at some point in his/her school studies. A Frege proof consists of lines that are propositional formulas built from propositional variables x_i and Boolean connectives $\neg$ (NOT), $\wedge$ (AND), $\vee$ (OR) or $\rightarrow$ (IMPLICATION). A Frege system then comprises a finite set of axiom schemes and inference rules. For example, $F \vee \neg F$ is a possible axiom scheme. A *Frege proof* of a Boolean formula L_ℓ from a set of formulas $\mathcal{F}$ is a sequence of formulas $(L_1, \ldots, L_\ell)$ where each formula is either in $\mathcal{F}$, a substitution instance of an axiom, or can be inferred from previous formulas by a valid inference rule, for example *modus ponens*:

$$\frac{F \quad F \rightarrow G}{G}. \tag{1.1}$$

Frege systems are required to be *sound* and *complete*, that is each tautology has to have a Frege proof (from $\mathcal{F} = \emptyset$) and no formula that is not a tautology should have a Frege proof (from $\mathcal{F} = \emptyset$). Or, equivalently, each contradiction has to have a Frege refutation, i.e., a proof of the trivially false formula $\bot$, and no formula that is not a contradiction should have a Frege refutation. The exact choice of the axiom schemes, inference rules and basis of Boolean connectives does not matter too much as long as the system remains sound and complete. Indeed, any two Frege systems are p-equivalent [CR79, Rec75] and [Kra95, Theorem 4.4.13]. A concrete example of a Frege system over the base of Boolean connectives $\{\vee, \neg\}$ is the following set of inference rules from [Sho67, p. 21]:

$$\frac{}{F \vee \neg F} \quad \frac{F}{F \vee G} \quad \frac{F \vee F}{F} \quad \frac{F \vee (G \vee H)}{(F \vee G) \vee H} \quad \frac{F \vee G \quad H \vee \neg G}{F \vee H}, \tag{1.2}$$

where F, G, H are Boolean formulas with logical connectives in $\{\vee, \neg\}$. The second rule listed in eq. (1.2) is called the *weakening* rule. The last of the rules listed in eq. (1.2) is the *cut* rule.

There are several common restrictions that can be imposed on Frege proofs. Recall that the logical depth of a Boolean formula is the maximum number of alternations of logical connectives in the formation tree of the formula in any path in that tree. For instance $(\neg(\neg x \vee y)) \vee \neg z$ is a formula of depth 3. Then fixing some depth d, the Frege system where we allow only proofs with lines (formulas) of depth at most d is called a *depth-d* Frege. An important example of depth-d Frege system is depth-1 Frege, also called *resolution* [Bla37, Rob65]. Resolution refutations are formulas of depth 1 over $\vee$ and $\neg$, that is each line is a clause. But now CNF formulas are formulas of logical depth 2 so in resolution CNF formulas are just considered as a set of clauses. This is a proof system we will see throughout this work and in particular in Chap. 2. Formally speaking, the set of inference rules of resolution is the one we saw in eq. (1.2) but now F, G, H are clauses. Let's focus for a moment on the cut rule. In resolution it must have the following form:

$$\frac{C \vee x \quad D \vee \neg x}{C \vee D} ,\qquad (1.3)$$

where C, D denote clauses and x is a variable that we say is *resolved*. Moreover it is easy to show that resolution and resolution *without* the weakening rule are p-equivalent.

To understand the complexity of resolution proofs, various hardness measures were defined and investigated. Historically, the first and most studied is the *size* of resolution proofs, that is the number of clauses in them. The very first lower bounds on the size of resolution proofs were proved by [Tse83, Hak85] and since then the complexity of resolution proofs was investigated a lot, see the surveys [Seg07, BP01, Pit11, Raz01, Pud08, Nor15]. First the interest in resolution was driven by the fact that lower bounds for this system were a first step towards lower bounds for higher-depth or unbounded-depth Frege systems. Nowadays the interest in resolution lies mostly in its connections to algorithms for deciding the satisfiability of CNF formulas, the so-called SAT-solvers, see Sect. 1.1.4.

Two general techniques that turned out to be very useful in proving lower bounds for resolution are the *feasible* interpolation [Kra97] and the *size-width* inequality, see eq. (2.10) and [BW01]. The feasible interpolation technique reduces the problem of proving size lower bounds in a given proof system to the problem of proving size lower bounds in some circuit class associated with the proof system. This technique applies for instance also to some geometric systems, see Sect. 1.1.3. The size-width inequality instead reduces the problem of proving size lower bounds in resolution to the easier task of proving width lower bounds, that is lower bounds on how large must be the largest clause in the proof. We will talk more in detail about the size-width inequality and a similar inequality that holds in an algebraic proof system respectively in Chap. 2 and Chap. 4.

So far we talked about resolution that is depth-1 Frege. Allowing more expressive lines we encounter first the propositional proof system called *resolution over k-DNFs* [Kra01], that is a Frege system with inference rules given again by eq. (1.2) but now each line in the proof is allowed to be a k-DNF formula, that is a Boolean formula in *Disjunctive Normal Form* (DNF).[2] For constant k, resolution over k-DNFs is even weaker than depth-2 Frege. It will become p-equivalent to it if, when refuting an unsatisfiable CNF formula in n variables, we allow lines that are $(\log n)^{O(1)}$-DNFs.

---

[2] A k-DNF is a disjunction of conjunctions of at most k variables or negated variables.

Allowing higher-depth formulas as lines in Frege we get first the constant-depth Frege systems. All of these are exponentially separated, that is depth-$(d+1)$ Frege is exponentially stronger than depth d Frege [Ajt94, PBI93, KPW95b]. Yet it is open whether such separation exists when both systems are just considered to refute unsatisfiable CNF formulas.

Then when we reach depth-$O(\log n)$ Frege we get that it is p-equivalent to unconstrained Frege. The strongest proof system for which we know that there are contradictions requiring exponentially long proofs is bounded-depth Frege, see [Kra94, KPW95a, PBI93]. Such lower bounds rely on *Switching Lemmas*, see [Hås87, Bea94]. We will not see such proofs directly but we will see an application of a version of the Switching Lemma in Chap. 8 for a subsystem of resolution, see Lemma 2.1. Recently Håstad showed a super-polynomial lower bound for depth-$o(\log n / \log \log n)$ Frege for some CNF formulas in n variables (*pers. comm.*), see Sect. 7.4 for more details on the formula used.

Proving any super-polynomial lower bound on the size of (unconstrained) Frege proofs is a major open problem in proof complexity. For example a class of formulas that is conjectured to be hard for Frege (and indeed in *any* propositional proof system) is the random k-CNF formulas, see Sect. 7.2.

Another major open problem in proof complexity is to prove exponential-size lower bounds for $AC_0[p]$-*Frege* [Pud08, Problem 10]. This is a Frege system where each line has only formulas of bounded depth but such formulas together with the usual Boolean connectives, say $\{\vee, \neg\}$, can also use a MOD_p connective. Semantically $MOD_p(x_1, \ldots, x_m) = 1$ if and only if $\sum_i x_i \equiv 0 \pmod{p}$ but formally we add to the inference rules in eq. (1.2) some new inference rules modeling the behaviour of MOD_p connectives.

TC_0-*Frege* is defined similarly, where the lines are formulas of bounded depth; say over the logical connectives $\{\vee, \neg\}$, but they also use some *threshold* connectives (and the inference rules define how those threshold connectives behave). This results in a system that p-simulates $AC_0[p]$-Frege and is p-simulated by (unconstrained) Frege. Proving exponential lower bounds for TC_0-Frege is also a big open problem, and seems even harder than proving lower bounds for $AC_0[p]$-Frege. More generally, given a circuit class $\mathcal{C}$, $\mathcal{C}$-*Frege* is a restriction of Frege where lines are circuits from the class $\mathcal{C}$; for a formal definition see [Jeř05].

It turns out that the study of algebraic proof systems can be seen as a first step towards size lower bounds for $AC_0[p]$-Frege and some geometric proof systems are special cases of TC_0-Frege proofs.

1.1.2 Algebraic Proof Systems

The idea of using propositional proof systems to capture basic algebraic facts and constructions dates back to [BIK⁺94] where a propositional proof system motivated by Hilbert's Nullstellensatz was introduced. Then [CEI96] introduced an even more natural algebraic proof system, *polynomial calculus*, that directly simulated the process of generating an ideal from a given set of generators. This proof system is the other proof system we will investigate in this book, in particular in Chap. 4. The algebraic proof system introduced in [CEI96] for some technical (somewhat non-relevant) reasons did not p-simulate resolution. It was then improved by [ABRW02] to a system they call polynomial calculus *with resolution* which is a minimal extension of both resolution and the system from [CEI96]. This technical difference relies on two ways, one more efficient than the other,[3] of encoding propositional Boolean formulas into polynomials, see Sect. 4.1. We will ignore this small issue for the moment and just call polynomial calculus what should more precisely be called polynomial calculus with resolution.

After encoding a CNF formula F as an equisatisfiable set of polynomials P_F (in some ring of polynomials over some field $\mathbb{F}$), we can show that F is unsatisfiable just by showing that the polynomial 1 is in the ideal generated by P_F. This is done through the following two inference rules

$$\frac{p \quad q}{\alpha p + \beta q} \qquad \frac{p}{xp} \ , \tag{1.4}$$

where p, q are polynomials, x is any variable and $\alpha, \beta \in \mathbb{F}$. That is we can perform arbitrary linear combinations of already inferred polynomials and we can multiply an inferred polynomial by a variable. These rules model the fact that ideals are closed under the previous two operations. Alternatively, we can show that $1 \in \langle p_1, \ldots, p_m \rangle$ in a static way, that is just exhibiting polynomials $q_1, \ldots, q_m$ such that

$$1 = \sum_{i=1}^{m} p_i q_i \ . \tag{1.5}$$

This is the *Nullstellensatz* proof system [BIK$^+$94] and it is p-simulated by polynomial calculus (if we consider polynomials in the two systems having coefficients in the same field $\mathbb{F}$).

In polynomial calculus the polynomials are handled in their expanded form as sums of monomials, and the size of a proof is measured as the total number of monomials appearing in it. The first example of formulas requiring exponentially long proofs in polynomial calculus was already given in [CEI96], and since then many other size lower bounds were proved, see for instance in [Raz98, IPS99, GL10b, MN15]. Indeed, [IPS99] showed that a generic way of proving size lower bounds in polynomial calculus it to prove degree lower bounds. That is there is a *degree-size* inequality analogous to the width-size inequality in resolution. A lot of results on the complexity of resolution proofs are indeed qualitatively similar to results on the complexity of polynomial calculus proofs. For instance a width upper bound of w implies a resolution size upper bound of $n^{O(w)}$, for CNF formulas in n variables. Similarly a degree upper bound of d implies a polynomial calculus

---

[3] The difference between the two encodings is that the latter one has separate formal variables to encode positive and negative literals over the same Boolean variable. Then, clauses with many literals are encoded more efficiently regardless of the polarity of the literals, which allows polynomial calculus with resolution to p-simulate resolution.

size upper bound of $n^{O(d)}$ for formulas over n variables [CEI96]. Both the upper bound in resolution and the one in polynomial calculus are tight [ALN14]. As for the size-width inequality in resolution, the degree-size inequality is essentially optimal, see respectively [BG01] and [GL10b]. Indeed the formulas used to show both results are the same: some combinatorial principles encoding the fact that in any finite partial ordering there must be a minimal element (*ordering principles*). Despite this analogy between the results for resolution and polynomial calculus the latter ones are usually more difficult and technically more involved, although sometimes they predate the analogous, easier, ones for resolution, for instance in the case of the size-degree and the size-width inequalities and for some of the space-related results we show in this book.

The main motivation to study polynomial calculus is the big open problem in proof complexity we mentioned before, that is proving exponential-size lower bounds for $AC_0[p]$-Frege. This is tightly connected to polynomial calculus and the Nullstellensatz proof system. Indeed the first introduction of algebraic proof systems in [CEI96] was already informally directed towards size lower bounds for $AC_0[p]$-Frege. Formally, the problem of proving size lower bounds in $AC_0[p]$-Frege was reduced to the problem of proving degree lower bounds in the Nullstellensatz proof system but for quite complicated sets of polynomials where some extension axioms might be used [BIK+97].

Secondarily another motivation to study polynomial calculus is that we can build SAT algorithms based on it, see Sect. 1.1.4. For instance, an algorithm might encode CNF formulas F as an equisatisfiable set of polynomials P_F and then use the Gröbner-basis algorithm to detect whether $1 \in \langle P_F \rangle$ and hence whether F is satisfiable or not. This is an algorithm whose running time, if F is unsatisfiable, is at least the shortest refutation of F in polynomial calculus. For the record this algorithm is not competitive with state-of-the-art SAT algorithms based on resolution.

Other types of algebraic proof systems have been considered in the literature, for example in [Pit96, GHP01, GH01, BIK+97, BGIP99, RT08, GP14]. A notable example is the *Sum-of-Squares* proof system [Ste73]. This is a static proof system based on a powerful modification of the Nullstellensatz proof system: a proof of the fact that a set of polynomials $\{p_1, \ldots, p_m\}$ say with coefficient in $\mathbb{R}$ is unsatisfiable is another set of polynomials $\{q_1, \ldots, q_m\} \cup \{s_1, \ldots, s_\ell\}$ such that

$$-1 = \sum_{i=1}^{m} p_i q_i + \sum_{j=1}^{\ell} s_i^2 \ . \tag{1.6}$$

Recent developments, see the survey [BS14], seem to indicate that SAT-solvers based on sum-of-squares might be serious rivals to resolution-based SAT-solvers, although it has to be mentioned that the most relevant size lower bound technique for sum-of-squares is still based on resolution width [Gri01].

1.1.3 Geometric Proof Systems

In this book we will not cover geometric proof systems in detail. Anyway it will be natural to talk about them in connection with SAT-solvers in Sect. 1.1.4 and regarding some recent results on their size and space complexity. So let's briefly introduce one of those systems: *cutting planes* [Gom63, Chv73]. This proof system captures some geometric methods used for integer linear programming: the so-called *Gomory-Chvátal cuts* to transform a polytope defined by a system of linear inequalities into its integral hull. Informally, if a system of linear inequalities has no integral solution then the inequalities define a polytope with empty integral hull and a sequence of Gomory-Chvátal cuts can be taken as a witness of the fact that there are no integral solutions. W. Cook in [CCT87] used this idea to define the *cutting planes* propositional proof system. Boolean formulas in CNF form are translated into a set of linear inequalities in a straightforward way preserving equisatisfiability. For example, the CNF formula $F = (x \vee \neg y \vee z) \wedge (\neg x \vee z)$ is translated into the following set of linear inequalities:

$$\{x + (1 - y) + z \geqslant 1, \ (1 - x) + z \geqslant 1, \ 0 \leqslant x \leqslant 1, \ 0 \leqslant y \leqslant 1, \ 0 \leqslant z \leqslant 1\} \, . \quad (1.7)$$

The original formula is satisfied if and only if the associated set of inequalities has an integral solution. Then a cutting planes refutation of an unsatisfiable CNF formula F is a sequence of linear inequalities $\sum_i a_i x_i \geqslant b$ with the $a_i \in \mathbb{Z}$ and $b \in \mathbb{R}$ and there are inference rules to take linear combinations of inequalities and to perform (a version of) Gomory-Chvátal cuts, that is we have the following inference rule:

$$\frac{\sum_i a_i x_i \geqslant b}{\sum_i a_i x_i \geqslant \lceil b \rceil} \, . \quad (1.8)$$

Notice that this inference rule is only sound if we just allow the x_is to take integral values. Cutting planes is a restricted version of TC_0-Frege, it p-simulates resolution and it is exponentially stronger than it. We know some exponential lower bounds on size based on a generalization of the interpolation technique, see [HC99, Pud97], but for instance we don't know whether random 3-CNF formulas require (as we expect) exponentially long refutations in cutting planes, see Sect. 7.2 for more details.

1.1.4 Connection with SAT-Solvers

The lower bounds shown in propositional proof complexity usually have an algorithmic counterpart. For some proof systems, this connection to algorithms is what drives the research about them nowadays. For example, resolution, proposed already in [Rob65] for automated theorem proving, is mostly studied due to its importance in applied contexts such as SAT-solvers, in particular due to a connection to the DPLL algorithm and the CDCL solvers. The *Davis-Putnam-Logemann-Loveland* (DPLL) [DP60, DLL62] algorithm is a backtracking method to search for assignments satisfying a CNF formula. It is a well-known result that the track of a run of the DPLL algorithm on unsatisfiable CNF formulas is equivalent to tree-like resolution, a subsystem of resolution where only proofs having a tree-like structure are allowed.

A strengthening of the DPLL algorithm was defined a a series of works, [BJS97, SS99, MMZ$^+$01], where the authors introduced the idea of *Conflict-Driven Clause Learning* (CDCL) as a way for DPLL-based SAT-solvers to cut the search space and avoid duplicated work. Informally, this is done by performing a *conflict analysis* when the search for an assignment leads to a contradiction and then *learning* a clause encoding a reason for that failure. By construction, resolution p-simulates CDCL solvers viewed as proof systems and, under certain assumptions on the behavior of the CDCL solver, the converse also holds [PD11, AFT11]. The crucial hypothesis in both [PD11, AFT11] is that the CDCL solver *never* deletes a learned clause. We stress that this is not a realistic hypothesis and, at the moment, it is not known whether CDCL solvers, under a more realistic modeling of the memory usage, p-simulate resolution.

Regarding algebraic proof systems, the original name for polynomial calculus in [CEI96] was *Gröbner* proof system due to its tight connection with the Gröbner-basis algorithm, and indeed the system was intended to be a potential candidate for efficient new SAT-solvers. There are SAT-solvers based on the Gröbner-basis algorithm such as PolyBoRi [BD09, BDG$^+$09] but they are not competitive from the point of view of performance with state-of-the-art CDCL solvers. Some, very limited, form of algebraic reasoning is starting to be integrated into CDCL solvers but, at the moment, this consists mostly of some form of Gaussian elimination. At the moment of writing, it seems that research about SAT-solving algorithms is not headed towards extending the CDCL paradigm with more algebraic techniques. It seems to be mainly headed in another direction: trying to build algorithms extending the CDCL paradigm by some geometric reasoning (*pseudo-Boolean* solvers), that is algorithms formalizable in some fragment of the cutting planes proof system that is stronger than resolution. For more details on the connection between proof complexity, in particular resolution and polynomial calculus, and SAT-solvers we refer to [Nor15]. For more details on pseudo-Boolean solvers we refer to [RM09, DGP04, DGLP04, DGH$^+$05].

The running time and the memory consumption of CDCL solvers are related to resolution size and resolution *space*. Lower bounds for the size and space complexity measures for resolution will translate into lower bounds on the running time and size of some auxiliary memory used by CDCL solvers. On the other hand, it is not known whether a generic CDCL solver p-simulates resolution on unsatisfiable CNF formulas and, from the space complexity point of view, it is perfectly possible that if the space usage of CDCL solvers is bounded, then they run, for example, in exponential time (or even worse) on instances easy for resolution. However, size and space are not the only measures that are interesting with respect to applications and the question of what constitutes a good hardness measure for practical SAT-solving is essentially open and a very important one from the practical point of view, see [BK14, ABLM08, JMNZ12].

1.2 Space of Proofs

The problem of the *space* taken by propositional proofs was posed for the first time by Armin Haken during the workshop *"Complexity Lower Bounds"* held at the Fields Institute, Toronto in 1998. Before that, apparently, the only paper investigating the space of proofs was [Koz77] but the author dealt only with equational theories involving no propositional connectives. The formal definition of the space taken by resolution proofs was given in [ET01] building on [KL94] and this definition was generalized later to other proof systems in [ABRW02].

Intuitively, the space required by a refutation is the amount of information we need to keep simultaneously in memory as we work through the proof and convince ourselves that the original propositional formula is unsatisfiable. This model is inspired by the definition of space complexity for Turing machines, where a machine is given a read-only input tape from which it can download parts of the input to the working memory as needed. In the literature this model is sometimes called the *blackboard* model. The name comes from seeing a proof as given by someone (a teacher) to some verifier (a class of students). The teacher want to show that a particular CNF formula is contradictory and he does this by writing down clauses and performing inferences on a given blackboard. Then the students verify his proof and in this analogy they understand inferences based on the rules of some particular proof system, for example Frege (or resolution or polynomial calculus).

As [ABRW02] point out, the very first question, when starting the investigation of space, is how to measure the memory content/blackboard size at any given moment in time for a specified propositional proof system. Recalling [Kra95], the most customary measures for the size complexity of propositional proofs are the bit size and the number of lines. Of the two the bit size is the more important and can be defined analogously also for space complexity. Similarly to what is done for size, usually we do not directly measure the bit size, but a logarithmically related measure that, in the case of space, is the total number of literals in memory, the *total space*.[4] Regarding the upper bounds, all contradictions can be refuted within polynomial space for any "reasonable" space measure, see [ET01].

The line complexity for strong enough proof systems, such as Frege, is not an adequate space measure. If the language of the proof system is strong enough to handle unbounded fan-in $\wedge$ gates, then just constant line space is sufficient as one can always use a big-$\wedge$ of all the formulas derived in previous steps. Moreover [ABRW02, Theorem 6.3] showed that any contradiction in n variables has a proof in Frege with total space $O(n)$. This fact somehow justifies the study of space for proof systems where super-linear lower bounds on space may be achieved (although total space in Frege is still a meaningful complexity measure).

---

[4] In [ABRW02] this measure is called *variable* space but we prefer to call it *total* space following [BN09, BN08, BN11, Nor09, Nor13, Urq11a]. The reason to do this is to distinguish this measure from another one where different occurrences of the same variable are not counted, see [Urq11a]. We call this latter one *variable* space, see Sect. 3.4.

Resolution, polynomial calculus and cutting planes are not closed under $\land$. In resolution the lines are just clauses and the line space, usually called *clause space*, makes perfect sense. This complexity measure was actually proposed in [ET01]. The line space makes sense also for stronger proof systems, such as polynomial calculus, where we consider the number of distinct *monomials* appearing in memory, *monomial space*, or cutting planes, where the number of linear inequalities in memory is considered, *inequality space*, see [GPT15]. The line space was also studied for resolution over k-DNFs [EGM04, BN09].

Unlike what happens for size, for some space measures the actual inference rules of the proof systems do not matter. That is the space lower bounds hold for some *semantic* version of the proof systems. What matters in such cases is the type of lines handled by the system, e.g., clauses or polynomials or generic Boolean formulas. This phenomenon was first observed in [ABRW02] for the clause space, for monomial space for a restricted class of formulas and for Frege in general, see [ABRW02, Corollary 6.6].

We end this very introductory part by recalling that space complexity has been studied also from the point of view of trade-offs, say between space and size. That is results showing that some formulas may have short proofs and proofs using small space but those two features cannot be achieved at the same time. In resolution trade-offs between clause space and size have been shown for instance in [BN08, NH13, BN11, Nor13, BNT13, BBI12]. In polynomial calculus trade-offs between monomial space and size have been studied for instance in [BNT13, Nor13]. Recently a new kind of trade-offs was also studied, namely *super-critical* trade-offs say between size and width or width and clause space, see respectively [Raz16a] and [BN16]. That is there are formulas that have both short resolution proofs and also low-width space proofs, but each proof using not too large width must have doubly exponential (tree-like) resolution refutations. Similarly for clause space and width: there are formulas that have both short resolution proofs and also low-width space proofs but each proof using not too large width must have clause space greatly exceeding the linear worst-case upper bound.

1.3 Summary of Results

From a very high-level point of view, the backbone of this book is the use of combinatorial families of assignments (and games) to prove lower bounds. In proof complexity game theoretic methods and combinatorial characterizations have a long history. This started from the very first exponential-size lower bound for resolution by [Hak85]. Then they have been widely used to characterize complexity measures, see for example [Pud00, AD08, BK14, BGL13, BGL10]. Here we apply those ideas to prove:

- Lower bounds for monomial space in polynomial calculus, see Chap. 4.
- Lower bounds for total space in resolution, see Chap. 3.
- Strong size lower bounds in (a subsystem of) resolution, see Chap. 8.

In Part I we collect the general techniques and results, Part II contains the applications of those techniques to notable families of CNF formulas and Part III is a postlude on resolution size lower bounds.

Regarding space in polynomial calculus the main results, in short, are the following:

- a new combinatorial framework to prove space lower bounds in polynomial calculus, see Theorem 4.2;
- asymptotically optimal lower bounds on the space needed to refute random k-CNF formulas (and the graph pigeonhole principle) in polynomial calculus, see Theorem 7.1 (and Theorem 7.4). This result was conjectured to be true and posed as an open problem in many works, see for instance [BS01, ABRW02, FLN$^+$15].

Regarding total space in resolution the main result is a general technique to prove *total space* lower bounds in resolution, see Theorem 3.6. Then, as corollaries, we have the following:

- An asymptotically optimal total space lower bound in resolution for *Tseitin formulas* over d-regular expander graphs, see Theorem 7.8. This result completely answers an open problem from [ABRW02, Open question 2].
- An asymptotically optimal total space lower bound in resolution for random k-CNF formulas, see Theorem 7.2. This result completely answers an open problem from [ABRW02, BS01, FLM$^+$13] among others.
- An optimal separation of resolution and *semantic* resolution from the point of view of the total space measure. This result completely answers [ABRW02, Open question 4] for resolution.

Regarding size and width in resolution we prove a *strong* width lower bound for resolution, see Theorem 8.1, and a *strong* size lower bound for a generalization of *regular* resolution, see Corollary 8.2.

All the results above are from my Ph.D. thesis [Bon15] and some earlier works [BG13, BGT14, BT15, BG15, BBG$^+$17, BT16a, BT16b, BGT16, Bon16]. Before those works there was only one work [ABRW02] proving some total space lower bounds in resolution. Regarding monomial space there were two works [ABRW02, FLM$^+$13] showing some monomial space lower bounds in polynomial calculus. All those results are now shown in Chap. 5 and Sect. 4.5.1 as applications of the general techniques we introduce here. More information on the history of the results shown in this book is at the end of each chapter in the **History** section. Each chapter contains also an **Open Problems** section on the open problems naturally arising from the chapter.

Part I
General Results and Techniques

Chapter 2
Resolution

In this chapter we recall some basic general facts about the proof complexity of *resolution* [Bla37, Rob65]. This will be a common background for the results in Chap. 3 and Chap. 8. First we describe the most common restrictions placed on the type of resolution refutations, that is regular and tree-like resolution refutations (Sect. 2.1), then we prove a non-trivial size upper bound (Sect. 2.2) and finally we move to the complexity measures known as width and asymmetric width (Sect. 2.3).

A *resolution derivation* of a clause C from a CNF formula F is a sequence of clauses $\pi = (C_1, \ldots, C_\ell)$ where $C_\ell = C$ and each C_i is either a clause in F or there are C_j, C_k with $j, k < i$ such that $\frac{C_j \quad C_k}{C_i}$ is a valid instance of the resolution inference rule:

$$\frac{A \vee x \quad B \vee \neg x}{A \vee B}, \tag{2.1}$$

where A, B are clauses and x is a variable. If $C = \bot$ then π is a *resolution refutation* of F. Recall that the *size* of a resolution refutation $\pi = (C_1, \ldots, C_\ell)$ is $S(\pi) = \ell$.

2.1 Subsystems of Resolution

With resolution refutations can be associated branching programs, see [Kra95], and (labeled) Directed Acyclic Graphs (DAGs). Formally we choose to associate a resolution derivation with a DAG using the notion of *witness function*. This extra formality might seem unnecessary now but it will be needed later in Sect. 2.3.

Definition 2.1 (Witness Function). Let F be an unsatisfiable CNF formula and let $\pi = (C_1, \ldots, C_\ell)$ be a resolution derivation of some clause C_ℓ from F. A function $\sigma : [\ell] \to \binom{[\ell]}{2} \cup \{\star\}$ is a *witness* of the fact that π is a valid resolution derivation from F if and only if for each $i \in [\ell]$

1. $\sigma(i) = \{j, k\}$ is such that $j < i$, $k < i$ and $\frac{C_j \quad C_k}{C_i}$ is a valid instance of the resolution inference rule;

© Springer International Publishing AG, part of Springer Nature 2017
I. Bonacina, *Space in Weak Propositional Proof Systems*,
https://doi.org/10.1007/978-3-319-73453-8_2

2. $\sigma(i) = \star$ is such that C_i is a clause in F.

Notice that given a resolution derivation π there are, in general, many possible witnesses. Then, given a derivation $\pi = (C_1, \ldots, C_\ell)$ and a witness σ we can define a (labeled) DAG representing the structure of π according to σ. This is a graph with vertices the set $[\ell]$ and edges $\{(i, j) : i, j \in [\ell]$ and $i \in \sigma(j)\}$. Then each vertex i in the graph has label C_i.[1] Then, if π has a witness function that turns it into a DAG that happens to be a tree, we say that π is a *tree-like* derivation. If π has a witness function that turns it into a DAG where in every path each variable is resolved at most once, we say that π is a *regular* resolution derivation. Before telling more on tree-like and regular resolution derivations let's see an example.

Example 2.1 ([HY87]). Consider the unsatisfiable CNF formula

$$F = (\neg x \vee a \vee b) \wedge (x \vee a \vee b) \wedge (\neg b \vee z) \wedge (\neg a \vee c) \wedge (x \vee \neg y) \wedge$$
$$\wedge (\neg x \vee \neg w) \wedge (\neg c \vee x \vee y) \wedge (\neg c \vee \neg x \vee w) \wedge (x \vee y \vee \neg z) \wedge (\neg x \vee w \vee \neg z) . \quad (2.2)$$

A DAG representing a resolution refutation of F is in Fig. 2.1.

This example was given in [HY87]. There the authors proved that each minimal-size resolution proof of this formula corresponds to a DAG where there is a path with a variable resolved twice. That is minimal-size refutations of F are not regular.

We will see more about tree-like and regular resolution proofs in Chap. 8. For the moment we just recall that tree-like resolution is exponentially weaker than regular resolution, which in turn is exponential weaker than resolution, see [BG99, Stå96, AJPU07, Urq11b]. That is there are CNF formulas F having polynomial-size resolution refutations but every regular resolution refutation has exponential size (in the size of the formula F). The same holds between regular and tree-like resolution.

2.2 Size

In this section we review some upper and lower bounds on resolution size.

Let's then start with a trivial upper bound. For every unsatisfiable CNF formula F in n variables there exists a tree-like resolution refutation π of F such that

$$S(\pi) \leqslant 2^{n+1} - 1 . \quad (2.3)$$

This inequality, for instance, can easily be proved by induction on the number of variables of F or using the equivalence between tree-like resolution refutations and binary decision trees, see Sect. 2.2. Interestingly eq. (2.3) is not the best upper bound we can get for k-CNF formulas.

Theorem 2.1 ([BT16a]). *For every unsatisfiable k-CNF formula F in n variables there exists a tree-like resolution refutation π of F such that*

---

[1] Notice that the same label might be repeated several times in the proof DAG.

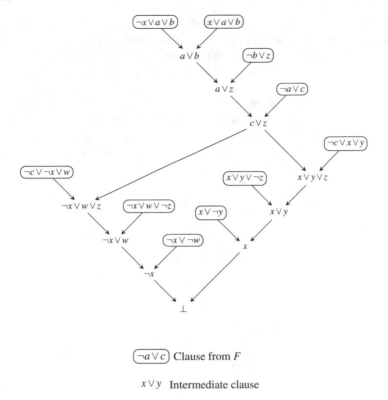

$\boxed{\neg a \vee c}$ Clause from F

$x \vee y$ Intermediate clause

Figure 2.1 An example of minimal-size resolution refutation of F

$$S(\pi) \leqslant 2^{\left(1-\Omega\left(k^{-1}\right)\right)n} . \tag{2.4}$$

To prove this result, instead of tree-like resolution refutations it is convenient to change the language a bit and talk of *decision trees*. Decision trees for an unsatisfiable CNF formula F are in a bijective correspondence with tree-like resolution refutations of F, see for instance [BGL13, Kra95].

Let F be an unsatisfiable CNF formula. A *decision tree* for F is a binary tree where the inner nodes are labeled with variables from the variables of F and the leaves are labeled with clauses from F. Each path in the decision tree corresponds to a Boolean assignment where a variable x gets the value 0 or 1 according to whether the path branches left or right at the node labeled with x. The condition on the tree is that each clause on the leaves is falsified by the Boolean assignment given by the path reaching the clause. The depth of a decision tree T is depth(T). Following [Bea94] we consider a particular type of decision trees: the *canonical* decision trees.

Definition 2.2 (Canonical Decision Trees). Given a CNF formula $F = \bigwedge_i C_i$ consider fixed orderings $\leqslant$ on the variables of F and $\preceq$ on the clauses of F. The *canonical* *decision tree* of F, $T(F)$, is inductively defined as follows: look at the first clause C of F according to the ordering $\preceq$ and let $F = C \wedge F'$. Then construct a full decision tree on the variables of C respecting the order $\leqslant$ of the variables, that is along each directed path from the root to the leaves the sequence of variables encountered $x_{i_1}, \ldots, x_{i_\ell}$ is such that $x_{i_1} \leqslant \cdots \leqslant x_{i_\ell}$. Each path from the root to a leaf defines a Boolean assignment and there is exactly one path from the root to a leaf v that correspond to a Boolean assignment that falsifies C. Label such a leaf v with the clause C. For all the other leaves w, let α_w be the Boolean assignment corresponding to the path from the root to the leaf w and replace the leaf w with $T(F'\!\restriction_{\alpha_w})$, see Fig. 2.2.

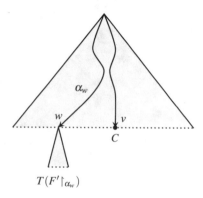

Figure 2.2 A canonical decision tree

Our upper bound on the size of tree-like resolution follows from the following non-trivial result on the depth of canonical decision trees. This result is in the spirit of the Switching Lemmas from [Hås87] although the statement is slightly different.

Lemma 2.1 (Håstad Switching Lemma [Bea94, Lemma 1]). *Let F be a k-CNF formula on n variables, d and ℓ be integers with $\ell \leqslant n/7$ and let α be an assignment chosen uniformly at random from the set of all Boolean assignments that have domain of size exactly $n - \ell$. Then*

$$\Pr_\alpha\left[\text{depth}(T(F\!\restriction_\alpha)) \geqslant d\right] \leqslant \left(\frac{7k\ell}{n}\right)^d . \quad \square \tag{2.5}$$

Given this result then the proof of the upper bound on the size of tree-like resolution refutations is indeed not difficult. The proof we give is modeled on [BT16a] and also inspired by [MRW05].

Proof (of Theorem 2.1). Let $\ell = n/14k$ and let $d = \ell/2$. By the Switching Lemma above, for a $1 - 2^{-d}$ fraction of partial assignments α with $|\text{dom}(\alpha)| = n - \ell$, the depth of $T(F\!\restriction_\alpha)$ is at most d. Then, by an averaging argument, there exists a subset

S of the variables of F with $|S| = n - \ell$ such for at least $1 - 2^{-d}$ of the partial assignments α with domain S, the depth of the canonical decision tree $T(F\restriction_\alpha)$ is at most d. Then we can construct a decision tree for F as follows: first we create a full decision tree on variables in S; then for each leaf with the corresponding restriction σ, we append $T(F\restriction_\sigma)$ to that leaf.

Then the number of leaves of this tree is upper bounded by

$$2^d 2^{n-\ell} + 2^{-d} 2^{n-\ell} 2^\ell, \tag{2.6}$$

since at most a 2^{-d} fraction of the leaves of the full decision tree on S can have maximal depth ℓ, see Fig. 2.3.

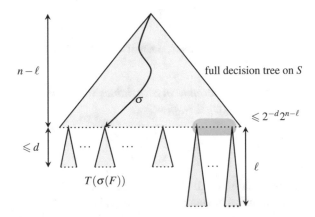

Figure 2.3 Size upper bound via canonical decision trees

Since $d = \ell/2$ then eq. (2.6) is upper bounded by

$$2^{n - \frac{\ell}{2} + 1} = 2^{\left(1 - \Omega\left(k^{-1}\right)\right)n}. \tag{2.7}$$

Hence we constructed a decision tree for F with size at most $2^{\left(1 - \Omega\left(k^{-1}\right)\right)n}$. Since, as already observed, decision trees correspond to tree-like resolution refutations we have the desired upper bound. $\square$

Regarding resolution size lower bounds it has been known for a long time now that there are unsatisfiable CNF formulas requiring exponential-size resolution refutations. All the formulas we will see in Part II are examples of such formulas. We refer to those chapters for some history on the proof complexity of such formulas. Here we just recall that the very first super-polynomial resolution size lower bound was obtained for *Tseitin formulas* [Tse83], see Sect. 7.4, and that nowadays we have many examples of formulas that need exponential-size resolution refutations. In Chap. 8 we will continue the investigation of resolution size, in particular of results matching the upper bound in eq. (2.4).

How do we prove size lower bounds then? There are some techniques tailored to specific classes of formulas—and we will see one such technique in Sect. 8.2—but here we want to focus on a very powerful and general technique to prove size (and space) lower bounds: the *width* method.

2.3 Width and Asymmetric Width

We introduce now two very helpful auxiliary complexity measures: the *width* and the *asymmetric width*. Given a sequence of clauses $\pi = (C_1, \dots, C_\ell)$, the *width* of π is

$$W(\pi) = \max_{i \in [\ell]} |C_i| . \tag{2.8}$$

The interest in this complexity measure is, for instance, the fact that it is tightly connected to size.

Theorem 2.2. *Let F be a CNF formula in n variables such that there exists a resolution refutation π of F with $W(\pi) \leqslant w$, then*

$$S(\pi) \leqslant \left(\frac{2en}{w} \right)^w . \tag{2.9}$$

Proof. This result is actually trivial since there are just at most $\sum_{i=0}^{w} \binom{2n}{i} \leqslant \left(\frac{2en}{w} \right)^w$ distinct clauses in n variables having at most w literals. $\square$

This trivial upper bound turns out to be tight: there are k-CNF formulas F in n variables refutable by resolution in width w but each resolution refutation of F must have size at least $n^{\Omega(w)}$ [ALN14].

The upper bound in the previous theorem is interesting because it upper bounds the running time of a simple algorithm that searches for resolution refutations. This algorithm just consists of deriving all resolution consequences of width $\leqslant w$ for increasing values of w until $\perp$ is derived [BS01].

The main interest in the width measure is actually due to the fact that width lower bounds imply size lower bounds.

Theorem 2.3 ([BW01, Corollary 3.4 and Theorem 3.5]). *Let F be an unsatisfiable k-CNF formula and suppose that every resolution refutation of F has width $\geqslant w$. Then for every resolution refutation π of F*

$$\log_2 S(\pi) = \Omega \left(\frac{(w-k)^2}{n} \right) , \tag{2.10}$$

and for every tree-like resolution refutation π' of F

$$\log_2 S(\pi') \geqslant w - k . \qquad \square \tag{2.11}$$

So, for instance, if for every resolution refutation π of a k-CNF formula F in n variables we have that $W(\pi) = \omega(\sqrt{n \log n} + k)$, then it immediately follows that every resolution refutation of F has super-polynomial (in n) size.

On the other hand, if a k-CNF formula F has resolution refutations of polynomial size then, again by eq. (2.10), it must have a resolution refutation π such that $W(\pi) = O(\sqrt{n \log n} + k)$. This doesn't necessarily mean that π has small— e.g., polynomial—size. Indeed, there are k-CNF formulas in n variables having polynomial-size resolution proofs, and hence having resolution refutations of width $O(\sqrt{n \log n} + k)$, but where this decrease in width comes at the expense of an exponential increase in the size of the proof [Tha14].

We say that there is a *trade-off* in resolution between two complexity measures if they cannot be optimized for the same resolution refutation. There is a vast literature on trade-offs; the interested reader might look at the introductory parts of [Nor15, Raz16a] or the chapter on resolution in [Juk12].

A relevant property of Theorem 2.3 is that the inequality in eq. (2.10) is "essentially" tight: there are 3-CNF formulas F_n in n variables having resolution refutations of polynomial size and with width $O(\sqrt{n})$ [BG01].

We now review another complexity measure, related to width, that we will need in Chap. 3. This is the *asymmetric* width [Kul00, Kul04]. Its definition is a bit more involved than that of width. Indeed to define it we need to be more precise about the DAG structure we associate with resolution proofs, and in particular its definition relies on the definition of the *witness* function of a refutation, see Definition 2.1.

Given a resolution refutation $\pi = (C_1, \ldots, C_\ell)$ of some CNF formula F and a witness function σ for π, the *asymmetric width* of C_i (with respect to π and σ) is

$$aw_{\pi,\sigma}(C_i) = \begin{cases} 0 & \text{if } \sigma(i) = \star \\ \min_{j \in \sigma(i)} |C_j| & \text{otherwise} . \end{cases} \qquad (2.12)$$

In other words, to determine $aw_{\pi,\sigma}(C_i)$ one has to look at the parents of C_i in the DAG associated with π by σ, take the minimum width of the parents and this value is $aw_{\pi,\sigma}(C_i)$. Then, similarly to what we did for the width, we define

$$aw_\sigma(\pi) = \max_{C_i \in \pi} aw_{\pi,\sigma}(C_i) . \qquad (2.13)$$

But, as already observed, a resolution refutation π can have many different witness functions and we want the *asymmetric* width to be independent of the chosen witness function. So we define the *asymmetric width* $aW(\pi)$ as the minimum of $aw_\sigma(\pi)$ over all the possible functions σ witnessing the validity of π, that is

$$aW(\pi) = \min_\sigma aw_\sigma(\pi) . \qquad (2.14)$$

Although width and asymmetric width have quite different definitions they are tightly connected complexity measures. Indeed, for every resolution refutation π it holds that $aW(\pi) \leqslant W(\pi)$ and the next lemma shows that a sort of converse of this fact holds too.

Lemma 2.2 ([Kul99, Lemma 8.5]). *Let F be an unsatisfiable k-CNF formula and suppose that every resolution refutation of F has width $\geqslant w$. Then for every resolution refutation π of F*

$$aW(\pi) + \max\{aW(\pi), k\} \geqslant w. \qquad (2.15)$$

The proof of this result is a bit technical and it is postponed till Sect. 2.3.2.

An interesting feature of the asymmetric width is that most of the results that hold for k-CNF formulas for the width also hold for the asymmetric width but without the dependency on k. Here we report the analogue of Theorem 2.3. In Chap. 3 we will see other examples concerning the space complexity of resolution.

Theorem 2.4 ([Kul04, Theorem 6.12]). *Let F be an unsatisfiable k-CNF formula and suppose that every resolution refutation of F has asymmetric width $\geqslant w$. Then for every resolution refutation π of F*

$$\ln S(\pi) \geqslant \frac{w^2}{8n}, \qquad (2.16)$$

and for every tree-like resolution refutation π' of F

$$\log_2 S(\pi') \geqslant w. \qquad \square \qquad (2.17)$$

The proof of this result is actually virtually the same as that of Theorem 2.3.

What we said so far might have convinced the reader that the width (and the asymmetric width) might be useful complexity measures to lower bound if we want to prove resolution size lower bounds. We have not yet addressed the question of *how* to prove width (or asymmetric width) lower bounds but this is what we do now.

There are essentially two general techniques to prove width lower bounds: one is via a "medium complexity clause" argument and the other uses a characterization of width lower bounds using families of Boolean assignments with certain combinatorial properties.

Suppose that we are given some CNF formula F and we want to prove that for every resolution refutation π of F we have that $W(\pi) \geqslant w$, for some parameter w. A "medium complexity clause" argument to prove this will have the following structure. To each clause C in π we associate a complexity measure $\mu(C)$—which of course will depend on the given formula F—such that for each clause $C \in F$, $\mu(C)$ is *small*, $\mu(\bot)$ is *large* and μ is sub-additive, that is for each pair of clauses C, C' and variable x,

$$\mu(C \vee C') \leqslant \mu(C \vee x) + \mu(C' \vee \neg x). \qquad (2.18)$$

Now these properties imply that there exists in π a clause C of intermediate complexity, i.e., such that $\mu(C)$ is not too small nor too large. Then, if we did things correctly, this clause C will have at least w literals. We will use this type of argument to prove a width lower bound in Sect. 8.3.

The other general way to prove width lower bounds is to use a characterization of width (and asymmetric width) lower bounds via families of Boolean assignments. Those characterizations are particularly helpful in proving space lower bounds, as we will see in Chap. 3.

2.3.1 Combinatorial Characterizations

We construct families of Boolean assignments with certain combinatorial properties that characterize width and asymmetric width lower bounds. We use standard notations in proof complexity for Boolean assignments, restrictions etc. The reader not familiar with these notations is invited to check the Notations on p. xv.

First we consider families of Boolean assignments characterizing width lower bounds. We call these families w-AD families from the initials of the authors introducing them: Atserias and Dalmau in [AD08].

Definition 2.3 (w-AD Families [AD08][2]). Given an unsatisfiable CNF formula F and $w \in \mathbb{N}$, we say that a family of Boolean assignments $\mathcal{F}$ is a w-AD family of Boolean assignments for F if the following properties hold:

1. $\mathcal{F}$ is non-empty;
2. for every $\alpha \in \mathcal{F}$ and every clause C in F, $C\!\restriction_\alpha \neq 0$ (**Consistency Property**);
3. if $\alpha \in \mathcal{F}$ and $\alpha' \subseteq \alpha$ is such that $|\mathrm{dom}(\alpha')| < w$, then for every variable $x \notin \mathrm{dom}(\alpha')$, there exists $\alpha'' \in \mathcal{F}$ with $\alpha'' \supseteq \alpha'$ such that $x \in \mathrm{dom}(\alpha'')$ (**Extension Property**).

Then a w-AD family of Boolean assignments for F characterize width lower bounds for F in the following sense.

Theorem 2.5 ([AD08, Theorem 2]). *Let F be an unsatisfiable CNF formula and w an integer. If for every resolution refutation π of F, $\mathrm{W}(\pi) > w$ then there exists a w-AD family for F. In the reverse direction if there exists a $(w+1)$-AD family for F then for every resolution refutation π of F, $\mathrm{W}(\pi) > w$.*

Proof. Suppose that for every resolution refutation π of F, $\mathrm{W}(\pi) > w$. We then want to construct a w-AD family of Boolean assignments $\mathcal{F}$ for F. A possible way to do this is as follows. Consider the set $\mathcal{C}$ of all clauses that have a resolution derivation from F of width at most w and take $\mathcal{F}$ to be the set of all Boolean assignments that don't falsify any clause in $\mathcal{C}$ or in F. More formally

$$\mathcal{F} = \{\alpha \; : \; \forall C \in \mathcal{C} \cup F, \; C\!\restriction_\alpha \neq 0\} \; . \tag{2.19}$$

By construction $\mathcal{F}$ clearly has the consistency property of Definition 2.3. Moreover, $\bot \notin \mathcal{C}$ hence the empty Boolean assignment λ is in $\mathcal{F}$ so $\mathcal{F}$ is non-empty. We just

---

[2] In [AD08] it is required that a w-AD family is closed under restrictions. This is indeed not needed.

have to show that if the extension property of Definition 2.3 does not hold then a contradiction will follow.

Suppose then that we have $\alpha \in \mathcal{F}$ and $\alpha' \subseteq \alpha$ such that $|\mathrm{dom}(\alpha')| < w$. By construction of $\mathcal{F}$ this implies that $\alpha' \in \mathcal{F}$. Moreover suppose that there is a variable $x \notin \mathrm{dom}(\alpha')$ such that for every $\alpha'' \supseteq \alpha'$ such that $x \in \mathrm{dom}(\alpha'')$ it holds that $\alpha'' \notin \mathcal{F}$. So, in particular $\alpha_0' = \alpha' \cup \{x = 0\}$ and $\alpha_1' = \alpha' \cup \{x = 1\}$ are not in $\mathcal{F}$. By construction then α_0' falsifies some clause C_0 in $\mathcal{C} \cup F$ and similarly α_1' falsifies some clause C_1 in $\mathcal{C} \cup F$. Since $\alpha' \in \mathcal{F}$ then it cannot falsify C_0 or C_1. This means that those clauses must be of the following form: $C_0 = C_0' \vee x$ with $C_0' \restriction_{\alpha'} = 0$ and similarly $C_1 = C_1' \vee \neg x$ and $C_1' \restriction_{\alpha'} = 0$. Since $|\mathrm{dom}(\alpha)| < w$ then $|C_0' \vee C_1'| < w$, $|C_0| \leqslant w$ and $|C_1| \leqslant w$. Then using the resolution rule we can derive $C_0' \vee C_1'$ from C_0 and C_1. From the fact observed before on the size of those clauses and the fact that C_0 and C_1 are in $\mathcal{C} \cup F$ then we have that $C_0' \vee C_1' \in \mathcal{C}$. But this is a contradiction since α' falsifies this clause and it is in $\mathcal{F}$.

Suppose now that there is a $(w+1)$-AD family $\mathcal{F}$ for F. We want to show that there is no resolution refutation π of F such that $W(\pi) \leqslant w$. For sake of contradiction let $\pi = (C_1, \ldots, C_\ell)$ be a resolution refutation of F such that $W(\pi) \leqslant w$. Since $\mathcal{F}$ is non-empty there is some Boolean assignment in it falsifying $C_\ell = \bot$. We then show that for each $i > 1$ there exists a Boolean assignment in $\mathcal{F}$ falsifying some clause $C_j \in \pi$ with $j < i$. This means that there exists some Boolean assignment in $\mathcal{F}$ falsifying C_1 but this is impossible since $C_1 \in F$ and no Boolean assignment in $\mathcal{F}$ can falsify clauses in F.

By induction suppose that we have a Boolean assignment $\alpha \in \mathcal{F}$ that falsifies $C_i \in \pi$. We then want to find a (possibly) new Boolean assignment in $\mathcal{F}$ that falsifies some $C_j \in \pi$ with $j < i$. By the consistency property of $\mathcal{F}$, $C_i \notin F$ so it is the resolvent of two previous clauses C_j and C_k in π. That is we have that $j < i$, $k < i$, $C_j = C \vee x$, $C_k = D \vee \neg x$ and $C_i = C \vee D$ for some clauses C, D and some variable x. If $x \in \mathrm{dom}(\alpha)$ then we are done, since α will falsify either C_j or C_k. Otherwise, by the assumption on the width of π, $|C \vee D| \leqslant w$. Consider $\alpha' \subseteq \alpha$ just assigning the variables in $C \vee D$. By assumption α' falsifies $C \vee D$ and can be extended to some Boolean assignment $\alpha' \in \mathcal{F}$ that assigns the variable x. This Boolean assignment α' then will either falsify C_j or C_k. $\square$

Similarly we have families of Boolean assignments characterizing asymmetric width lower bounds. We call these w-BK families from the initials of the authors introducing them: Beyersdorff and Kullmann in [BK14].

Definition 2.4 (w-BK Families [BK14]). Given an unsatisfiable CNF formula F and $w \in \mathbb{N}$, we say that a family of Boolean assignments $\mathcal{F}$ is a w-BK family of Boolean assignments for F if the following properties hold:

1. $\mathcal{F}$ is non-empty;
2. for every $\alpha \in \mathcal{F}$ and every clause C in F, $C \restriction_\alpha \neq 0$ (**Consistency Property**);
3. if $\alpha \in \mathcal{F}$ and $\alpha' \subseteq \alpha$ is such that $|\mathrm{dom}(\alpha')| < w$, then for every variable $x \notin \mathrm{dom}(\alpha)$ there exist $\alpha_0'', \alpha_1'' \in \mathcal{F}$ with $\alpha' \subseteq \alpha_0'$ and $\alpha' \subseteq \alpha_1'$ such that $\alpha_0'(x) = 0$ and $\alpha_1'(x) = 1$ (**Extension Property**).

Then a w-BK family of Boolean assignments for F characterize asymmetric width lower bounds for F in the following sense.

Theorem 2.6 ([BK14, Theorem 22]). *Let F be an unsatisfiable CNF formula and w an integer. The following two properties are equivalent:*

1. *for every resolution refutation π of F, $\mathrm{aW}(\pi) > w$;*
2. *there exists a w-BK family for F.*

For completeness we give a proof of this result although this proof can be seen as a modification of the proof of Theorem 2.5.

Proof. Suppose that for every resolution refutation π of F, $\mathrm{aW}(\pi) > w$ and let $\mathcal{C}$ be the set of clauses derivable from F with a resolution derivation of asymmetric width at most w. Let $\mathcal{F}$ be the family of all the Boolean assignments of maximal domain that do not falsify any clause in $\mathcal{C}$. Since $F \subseteq \mathcal{C}$ and $\bot \notin \mathcal{C}$ by hypothesis, then $\mathcal{F}$ is consistent and non-empty. We have just to show the *extension* property of $\mathcal{F}$: let $\alpha \in \mathcal{F}$, $\beta \subseteq \alpha$ such that $|\mathrm{dom}(\beta)| < w$ and $x \notin \mathrm{dom}(\alpha)$. For ease of notation, given $b \in \{0,1\}$ let

$$x^b = \begin{cases} x & \text{if } b = 0 \,, \\ \neg x & \text{if } b = 1 \,. \end{cases} \tag{2.20}$$

By the maximality of α we have that for each $b \in \{0,1\}$ there exists a clause C_b in $\mathcal{C}$ such that $\alpha_b = \alpha \cup \{x = b\}$ falsifies C_b. Since $\alpha \in \mathcal{F}$ then $x \in \mathrm{var}(C_b)$ and it must be that $C_b = C_b' \vee x^b$ where C_b' is a clause such that $\mathrm{var}(C_b') \subseteq \mathrm{dom}(\alpha)$ and $C_b' \restriction \alpha = 0$. Suppose, for sake of contradiction, that there exists $b \in \{0,1\}$ such that there is no $\beta' \in \mathcal{F}$ such that $\beta' \supseteq \beta$ and $\beta'(x) = b$. In particular $\beta_b = \beta \cup \{x = b\}$ is not in $\mathcal{F}$. Then, by construction, there exists a clause $D \in \mathcal{C}$ such that $D \restriction_{\beta_b} = 0$. Since $|\mathrm{dom}(\beta_b)| = |\mathrm{dom}(\beta)| + 1 \leqslant w$ then $|D| \leqslant w$. Moreover, since $\beta \subseteq \alpha$ and $\alpha \in \mathcal{F}$ does not falsify any clause in $\mathcal{C}$, then it must be that $D = D' \vee x^b$ and D' is a clause such that $D' \restriction_\alpha = D' \restriction_\beta = 0$. But now

$$\frac{D \qquad C_{1-b}}{D' \vee C_{1-b}'} \tag{2.21}$$

is a valid instance of the resolution rule. Hence, let π_D be a resolution derivation of D of minimum asymmetric width and similarly let $\pi_{D_{1-b}}$ be a resolution derivation of C_{1-b} of minimum asymmetric width. Then $\pi = \pi_D \circ \pi_{C_{1-b}} \circ (D' \vee C_{1-b}')$, the concatenation of π_D, $\pi_{C_{1-b}}$ and $(D' \vee C_{1-b}')$, is a resolution derivation of $D' \vee C_{1-b}'$ from F and by definition of asymmetric width,

$$\mathrm{aW}(\pi) \leqslant \max\{\mathrm{aW}(\pi_D), \mathrm{aW}(\pi_{C_{1-b}'}), \mathrm{aW}(D' \vee C_{1-b}')\} \leqslant w \,. \tag{2.22}$$

Hence $D' \vee C_{1-b}' \in \mathcal{C}$. On the other hand $D' \vee C_{1-b}' \restriction_\alpha = 0$ contradicting the fact that $\alpha \in \mathcal{F}$.

Suppose now that there is a w-BK family $\mathcal{F}$ and, for sake of contradiction, suppose that there is a resolution refutation $\pi = (C_1, \dots, C_\ell)$ of F such that $\mathrm{aW}(\pi) \leqslant w$. We proceed as in the proof of Theorem 2.5. Since $\mathcal{F}$ is non-empty, there is some Boolean assignment in it falsifying $C_\ell = \bot$. We then show that for each $i > 1$ there exists a Boolean assignment in $\mathcal{F}$ falsifying some clause $C_j \in \pi$ with $j < i$. This means that there exists some Boolean assignment in $\mathcal{F}$ falsifying C_1 but this is impossible since $C_1 \in F$ and no Boolean assignment in $\mathcal{F}$ can falsify clauses in F.

By induction suppose that we have a Boolean assignment $\alpha \in \mathcal{F}$ that falsifies $C_i \in \pi$. We then want to find a (possibly) new Boolean assignment in $\mathcal{F}$ that falsifies some $C_j \in \pi$ with $j < i$. By the consistency property of $\mathcal{F}$, $C_j \notin F$ so it is the resolvent of two previous clauses C_j and C_k in π. Suppose C_j and C_k are resolved on some variable x. If $x \in \mathrm{dom}(\alpha)$ then we are done, since α will falsify either C_j or C_k. Otherwise, by the assumption on the asymmetric width of π, at least one of C_j and C_k has width at most w and $x \notin \mathrm{dom}(\alpha)$. W.l.o.g. suppose that $|C_j| \leqslant w$ and consider $\alpha' \subseteq \alpha$ just assigning the variables in C_j. Then $|\mathrm{dom}(\alpha')| < w$ and by the extension property of $\mathcal{F}$ there are two extensions α'_0 and α'_1 of α' in $\mathcal{F}$, one setting $x = 0$ and the other setting $x = 1$ (possibly among other variables). Either α'_0 or α'_1 will falsify C_j. $\quad\square$

2.3.2 The "Equivalence" of Width and Asymmetric Width

We conclude this chapter with a self-contained proof of Lemma 2.2, restated below for the convenience of the reader. The proof we give is based on [BK14].

Restated Lemma 2.2 ([Kul99, Lemma 8.5]) *Let F be an unsatisfiable k-CNF formula and suppose that every resolution refutation of F has width $\geqslant w$. Then for every resolution refutation π of F*

$$\mathrm{aW}(\pi) + \max\{\mathrm{aW}(\pi), k\} \geqslant w . \tag{2.15}$$

Proof (of Lemma 2.2). Given a set of clauses A, an *A-input resolution derivation* of a clause C is a resolution derivation of C from A such that each application of the inference rule has at least one premise from A. The main property of A-input resolution derivations is the following: if each clause in A has at most k' literals and there exists an A-input derivation of a clause C then there exists a resolution derivation π of C from A such that

$$\mathrm{W}(\pi) \leqslant |C| + k' . \tag{2.23}$$

Notice that to prove eq. (2.23), it is sufficient to consider A-input *refutations*, that is A-input derivations of the empty clause $\bot$. Indeed, suppose we have an A-input derivation π of a clause C, and let α be the Boolean assignment of minimal domain mapping C to 0. Clearly $|\mathrm{dom}(\alpha)| \leqslant |C|$ and $\pi\restriction_\alpha$ is an $A\restriction_\alpha$-input refutation, hence, if the property we want to prove holds for input refutations, then there exists a resolution refutation π' of $A\restriction_\alpha$ such that $\mathrm{W}(\pi') \leqslant k'$. Then, by weakening the clauses in π' by literals in C and the fact that α is removing at most $|C|$ literals from each clause, we get a resolution derivation π'' of C from A such that $W(\pi'') \leqslant k' + |C|$.

So let's prove eq. (2.23) in the case when $C = \bot$ and there exists an A-input resolution refutation. Let $\mathcal{A}$ be the set of all k'-clauses A that have an A-input resolution refutation but for every resolution refutation π of A it holds that $W(\pi) > k'$. For sake of contradiction suppose that $\mathcal{A}$ is non-empty, so there will be some $\bar{A} \in \mathcal{A}$ with a minimum number of variables. Since $\bar{A} \in \mathcal{A}$ then it must be that $\bar{A}$ is non-trivial, i. e., $\bot$ cannot appear in $\bar{A}$. By hypothesis then we have an $\bar{A}$-input refutation π and let ℓ be the last literal resolved in π. Since π is an $\bar{A}$-input refutation it must be that either $\ell \in \bar{A}$ or $\neg\ell \in \bar{A}$. Without loss of generality suppose that $\neg\ell \in \bar{A}$. Now consider $\pi\restriction_{\ell=0}$. This is an $\bar{A}\restriction_{\ell=0}$-input resolution refutation and $\bar{A}\restriction_{\ell=0}$ has strictly fewer variables than $\bar{A}$; hence, by the minimality of $\bar{A}$, it cannot be in $\mathcal{A}$. So there exists some π' that is a refutation of $\bar{A}\restriction_{\ell=0}$ with $W(\pi') \leqslant k'$. But now

$$\pi'' = \bar{A} \circ \pi' , \tag{2.24}$$

the concatenation of the clauses in $\bar{A}$ and π', is a resolution refutation of $\bar{A}$. Then clearly $W(\pi'') \leqslant k'$, contradiction the fact that $\bar{A} \in \mathcal{A}$. Notice that π'' is not, in general, a valid $\bar{A}$-input resolution refutation. Yet it is a valid resolution refutation of $\bar{A}$, because $\neg\ell \in \bar{A}$ and hence each clause in $\bar{A}\restriction_{\ell=0}$ can be seen as the result of an inference step between some clause in $\bar{A}$ and $\neg\ell$.

Suppose now we are given a resolution refutation π of the k-CNF formula F and let $r = aW(\pi)$. Consider the set of clauses S defined as the closure of F under input derivations, that is

$$\begin{cases} S_0 & = F , \\ S_{i+1} & = S_i \cup \{C \text{ clause} : |C| \leqslant r \text{ and } C \text{ has an } S_i\text{-input resolution derivation}\} , \\ S & = \bigcup_i S_i . \end{cases}$$
$$\tag{2.25}$$

Notice that each clause in S has width at most $\max\{r,k\}$ and hence S is just a finite union as S_{i+1} can be strictly bigger than S_i at most $O\left(n^{\max\{r,k\}}\right)$ many times, since this is the number of clauses in n variables of width at most $\max\{r,k\}$. Now we claim to have the following two properties:

1. $\bot$ has an S-input resolution derivation;
2. if C has an S-input resolution derivation then there exists a resolution derivation π' of C from F such that

$$W(\pi') \leqslant r + \max\{r,k\} . \tag{2.26}$$

From the two previous properties we immediately get eq. (2.15).

To prove item 1, we just show that π is an S-input resolution refutation. For sake of contradiction, let C be the first clause in π inferred from previous C', C'' in π with both $C', C'' \notin S$. Since $aW(\pi) = r$ without loss of generality we have that $|C'| \leqslant r$, hence it must be that for each i, C' does not have an S_i-input resolution derivation, otherwise $C' \in S_{i+1}$ but we are supposing that $C' \notin S$. Hence, C' doesn't have an S-input resolution derivation either, contradicting the minimality of C in π.

We prove eq. (2.26) in item 2 by induction on S_i. For S_0 eq. (2.26) is clearly true. For the inductive step let C be a clause in $S_{i+1} \setminus S_i$ and let $S_i = \{C_1, \ldots, C_m\}$ and by assumption C has an S_i-input resolution derivation. By what we observed before in eq. (2.23) there exists some $\tilde{\pi}$ that is a resolution derivation of C from S_i such that

$$\mathrm{W}(\tilde{\pi}) \leqslant |C| + \max_{i \in [m]} |C_i| \leqslant r + \max\{r,k\} . \tag{2.27}$$

Finally, by the induction hypothesis, for each $i \in [m]$, C_i has a resolution derivation π_i from F of width at most $r + \max\{r,k\}$. Hence

$$\pi' = \pi_1 \circ \cdots \circ \pi_m \circ \tilde{\pi} \tag{2.28}$$

is a resolution derivation of C from F and

$$\mathrm{W}(\pi') = \max\{\mathrm{W}(\pi_1), \ldots, \mathrm{W}(\pi_m), \mathrm{W}(\tilde{\pi})\} \leqslant r + \max\{r,k\} . \qquad \square \tag{2.29}$$

2.4 Open Problems

There are no real open questions arising from this introductory chapter, except maybe one concerning the tightness of the inequality between width and asymmetric width in Lemma 2.2. A possible way to strengthen it, or to at least give a completely different proof, could be to use the characterizations of width and asymmetric width in terms of w-AD and w-BK families.

History

Theorem 2.1 might be folklore but anyway a proof, a variation of the one we show, was given in [BT16a]. For more information on and history of the asymmetric width we refer to [BK14].

Chapter 3
Space in Resolution

In this chapter we investigate the space complexity of resolution in particular from the point of view of the total space measure, see Sect. 3.3. We briefly review results about the clause space (Sect. 3.2) and the variable space measures (Sect. 3.4). We prove a general inequality between the total space measure and width, Theorem 3.6. Then, when talking about space it is natural to introduce a *semantic* version of resolution, see Sect. 3.1. We show the separation of resolution and semantic resolution and also a technique to prove total space lower bounds in semantic resolution, Theorem 3.7, and a bounded version of it, Theorem 3.8.

To formally define the space complexity of resolution it is convenient to change slightly the model of resolution derivations. In this new model, formalized in [ET01, ABRW02], a resolution derivation is a sequence of sets of clauses—called *memory configurations*—and an inference step is allowed to happen only between consecutive sets of clauses. To avoid confusion we call this new model *space-aware* resolution derivations.

More formally, a *space-aware* resolution refutation of a CNF formula F is a sequence $\pi = (\mathfrak{M}_0, \ldots, \mathfrak{M}_\ell)$ of sets of clauses, where $\mathfrak{M}_0$ is the empty set, $\mathfrak{M}_\ell$ contains the empty clause $\bot$, and each $\mathfrak{M}_{i+1}$ is derived from $\mathfrak{M}_i$ in one of the following three ways:

- $\mathfrak{M}_{i+1} = \mathfrak{M}_i \cup \{C\}$, where C is a clause from F (**Axiom Download**);
- $\mathfrak{M}_{i+1} \subseteq \mathfrak{M}_i$ (**Erasure**);
- $\mathfrak{M}_{i+1} = \mathfrak{M}_i \cup \{C\}$ where C follows from some clauses in $\mathfrak{M}_i$ by the resolution rule (**Inference**).

The size of a space-aware resolution derivation $\pi = (\mathfrak{M}_0, \ldots, \mathfrak{M}_\ell)$ is just $\sum_{i \in [\ell]} |\mathfrak{M}_i|$. It is then immediate to see that resolution and space-aware resolution are p-equivalent. Indeed given any CNF formula F and a resolution refutation of F of size s it is immediate to construct a space-aware resolution refutation of F of size $O(s^2)$.

Given a space-aware resolution refutation $\pi = (\mathfrak{M}_0, \ldots, \mathfrak{M}_\ell)$ we have three natural ways of measuring the space of π according to different ways of assigning to each $\mathfrak{M}_i$ how *spacious* it is. Indeed the following definitions will be sound for any

© Springer International Publishing AG, part of Springer Nature 2017
I. Bonacina, *Space in Weak Propositional Proof Systems*,
https://doi.org/10.1007/978-3-319-73453-8_3

sequence $\pi = (\mathfrak{M}_0, \ldots, \mathfrak{M}_\ell)$ of sets of clauses, regardless of whether it is a valid space-aware resolution refutation or not.

The *clause space* of $\pi = (\mathfrak{M}_0, \ldots, \mathfrak{M}_\ell)$ is

$$\mathrm{CSp}(\pi) = \max_{i \in [\ell]} |\{C \in \mathfrak{M}_i \; : \; C \neq \bot\}| \; . \tag{3.1}$$

The space of each $\mathfrak{M}_i$ is accounted only with respect to how many non-trivial clauses it has, regardless of the number of literals in each of those clauses, hence the name *clause space*.

The *total space* of $\pi = (\mathfrak{M}_0, \ldots, \mathfrak{M}_\ell)$ is

$$\mathrm{TSp}(\pi) = \max_{i \in [\ell]} \sum_{C \in \mathfrak{M}_i} |C| \; . \tag{3.2}$$

Now the space of each $\mathfrak{M}_i$ is measured in terms of the total number of instances of variables occurring in it, hence the name *total space*. If we don't want to double count variables appearing multiple times in $\mathfrak{M}_i$ we have the notion of *variable space*:

$$\mathrm{VSp}(\pi) = \max_{i \in [\ell]} \left| \bigcup_{C \in \mathfrak{M}_i} \mathrm{vars}\,(C) \right| \; . \tag{3.3}$$

Clearly for every $\pi = (\mathfrak{M}_0, \ldots, \mathfrak{M}_\ell)$ we have that

$$\mathrm{CSp}(\pi) \leqslant \mathrm{TSp}(\pi) \; , \tag{3.4}$$

$$\mathrm{VSp}(\pi) \leqslant \mathrm{TSp}(\pi) \; . \tag{3.5}$$

We will further investigate the clause space in Sect. 3.2, the total space in Sect. 3.3 and (briefly) the variable space in Sect. 3.4. For some of the space results we are going to see the actual inference rule of resolution is not relevant and hence we introduce, following [ABRW02], the notion of *semantic* space-aware resolution derivations.

3.1 Semantic Resolution

A *semantic* space-aware resolution refutation of a CNF formula F is a sequence $\pi = (\mathfrak{M}_0, \ldots, \mathfrak{M}_\ell)$ of sets of clauses, where $\mathfrak{M}_0$ is the empty set, $\mathfrak{M}_\ell$ contains the empty clause $\bot$, and each $\mathfrak{M}_{i+1}$ is a subset of $\mathfrak{M}_i \cup C$, where either

- C is a clause from F (**Axiom Download**); or
- C is implied by $\mathfrak{M}_i$, that is for each Boolean assignment α, if $\alpha \vDash \mathfrak{M}_i$ then $\alpha \vDash C$ (**Semantic Inference**).

Similarly to space-aware resolution refutations, also semantic space-aware resolution refutations can be analyzed from the point of view of the clause space, variable space and total space complexity.

We will also see a bounded version of semantic space-aware resolution refutations, that is d-semantic space-aware resolution refutations.

A d-*semantic* space-aware resolution refutation of a CNF formula F is a sequence $\pi = (\mathfrak{M}_0, \ldots, \mathfrak{M}_\ell)$ of sets of clauses, where $\mathfrak{M}_0$ is the empty set, $\mathfrak{M}_\ell$ contains the empty clause $\bot$, and each $\mathfrak{M}_{i+1}$ is a subset of $\mathfrak{M}_i \cup C$, where either

- C is a clause from F (**Axiom Download**); or
- C is implied by a set of at most d clauses in $\mathfrak{M}_i$, that is there exists $S \subseteq \mathfrak{M}_i$ with $|S| \leqslant d$ such that for each Boolean assignment α, if $\alpha \models S$ then $\alpha \models C$ (d-**Semantic Inference**).

Similarly to what we have seen before, we can easily adapt the space measures definitions to d-semantic resolution.

For each of these complexity measures we show some inequalities between the space complexity of space-aware and semantic space-aware resolution refutations and we also show some unconditional lower bounds. The unconditional lower bounds will hold for *semiwide* formulas. These are, very informally, formulas with some clauses with many literals while the remaining clauses might have few literals but they are "highly" satisfiable. Formally the definition is the following.

Definition 3.1 (Semiwide Formulas [ABRW02]). Given a CNF formula F' and a Boolean assignment α, we say that α is F'-*consistent* if α can be extended to some α' that satisfies F'. A CNF formula F is r-*semiwide* if $F = F' \wedge W$, where F' is a satisfiable CNF formula, and for each F'-consistent Boolean assignment α and each clause C from W, if $|\mathrm{dom}(\alpha)| < r$ then α can be extended to a F'-consistent Boolean assignment which satisfies C.

Some examples of n-semiwide formulas are the complete tree formulas, CT_n, see Sect. 4.1.1, and the pigeonhole principles PHP_n^m (and some of its variations), see Sect. 5.1.

3.2 Clause Space

Regarding the clause space we just prove two inequalities, an upper bound and a generic way to prove lower bounds. The reader interested in trade-offs between clause space and size can look at [BN08, NH13, BN11, Nor13, BNT13, BBI12, BN16].

Theorem 3.1 ([ET01, Theorem 2.1]). *For every unsatisfiable CNF formula F in n variables there exists π a space-aware refutation of F such that*

$$\mathrm{CSp}(\pi) \leqslant n+1 . \tag{3.6}$$

Proof. This result can be proven by induction on n. If $n = 1$, the formula F in the variable $\{x\}$ contain as a sub-formula $x \wedge \neg x$. Then

$$\pi = (\emptyset, \{x\}, \{x, \neg x\}, \{x, \neg x, \bot\}) , \tag{3.7}$$

is a space-aware refutation of F of clause space 2. For the inductive case, given a variable x in F consider $F\!\restriction_{x=0}$ and $F\!\restriction_{x=1}$. By the inductive hypothesis there exists a space-aware resolution refutation π'_x of $F\!\restriction_{x=0}$ with $\mathrm{CSp}(\pi'_x) \leqslant n$ and a space-aware resolution refutation $\pi'_{\neg x}$ of $F\!\restriction_{x=1}$ with $\mathrm{CSp}(\pi'_{\neg x}) \leqslant n$. If needed we can add to clauses in π'_x the literal x to obtain a space-aware derivation $\pi_x = (\mathfrak{M}_0, \dots \mathfrak{M}_\ell)$ of x from F in clause space at most n. Similarly for $\neg x$: we get a space-aware resolution derivation $\pi_{\neg x} = (\mathfrak{M}_{\ell+1}, \dots, \mathfrak{M}_t)$ of $\neg x$ from F with clause space at most n. Then

$$\pi = (\mathfrak{M}_0, \dots, \mathfrak{M}_\ell, \{x\}, \{x\}\cup\mathfrak{M}_{\ell+1}, \dots, \{x\}\cup\mathfrak{M}_t, \{x, \neg x\}, \{x, \neg x, \bot\}) \quad (3.8)$$

is a space-aware resolution refutation of F of clause space at most $n+1$. $\square$

Regarding the lower bounds we have some generic inequalities between clause space and (asymmetric) width. These rely on the following simple lemma.

Lemma 3.1 (Locality Lemma for Resolution). *Given a set of clauses A and a Boolean assignment α such that $\alpha \vDash A$, there exists $\beta \subseteq \alpha$ such that $|\mathrm{dom}(\beta)| \leqslant |A|$ and $\beta \vDash A$.*

Proof. For each clause $C \in A$ there is at least one literal ℓ_C in C such that $\alpha(\ell_C) = 1$. Take one such literal ℓ_C for each clause; clearly $|\{\ell_C : C \in A\}| \leqslant |A|$ and hence to satisfy A it is sufficient to restrict α to the set of variables appearing in the set of literals $\{\ell_C : C \in A\}$. $\square$

Theorem 3.2 ([AD08, BK14]). *Let F be an unsatisfiable k-CNF formula and suppose that every resolution refutation of F requires width $> w$ and asymmetric width $> w'$. Then for every resolution refutation π of F*

$$\mathrm{CSp}(\pi) > w - k\,, \qquad\qquad\qquad (3.9)$$
$$\mathrm{CSp}(\pi) > w'\,. \qquad\qquad\qquad (3.10)$$

Proof. Let's start with the second inequality. By Theorem 2.6, there exists a w'-BK family $\mathcal{F}$ for F. Suppose for sake of contradiction that there is a space-aware resolution refutation π of F such that $\mathrm{CSp}(\pi) \leqslant w'$. Let $\pi = (\mathfrak{M}_0, \dots, \mathfrak{M}_\ell)$. We show, by induction on i, that for every $\mathfrak{M}_i$ there exists some $\alpha_i \in \mathcal{F}$ such that $\alpha_i \vDash \mathfrak{M}_i$, that is for each clause $C \in \mathfrak{M}_i$, $C\!\restriction_{\alpha_i} = 1$. At stage $i = \ell$ we will get a contradiction since $\bot \in \mathfrak{M}_\ell$. The base case $i = 0$ is trivial: $\mathfrak{M}_0 = \emptyset$ and $\mathcal{F}$ is non-empty so take any Boolean assignment in $\mathcal{F}$ as α_0. For the inductive step if $\mathfrak{M}_{i+1}$ is obtained from $\mathfrak{M}_i$ by an erasure or an inference just set $\alpha_{i+1} = \alpha_i$. If $\mathfrak{M}_{i+1} = \mathfrak{M}_i \cup \{C\}$ with $C \in F$ we have two possibilities: either $\alpha_i \in \mathcal{F}$ assigns all variables in C or there is some variable x in C but not in $\mathrm{dom}(\alpha_i)$. In the first case, by the consistency property of $\mathcal{F}$, $C\!\restriction_{\alpha_i} = 1$ and hence we can just take $\alpha_{i+1} = \alpha_i$. In the second case we must have that $|\mathfrak{M}_i| < w'$. By Lemma 3.1 with parameters $A = \mathfrak{M}_i$ and $\alpha = \alpha_i$, there exists some $\beta \subseteq \alpha_i$ such that $|\mathrm{dom}(\beta)| \leqslant |\mathfrak{M}_i| < w'$ and $\beta \vDash \mathfrak{M}_i$. Then, by the extension property of $\mathcal{F}$ there are two Boolean assignments β_0, $\beta_1 \in \mathcal{F}$ extending β and setting x respectively to 0 and to 1. If the literal x is in C take $\alpha_{i+1} = \beta_1$; if the literal $\neg x$ is in C take $\alpha_{i+1} = \beta_0$. It is straightforward to check that in both cases $\alpha_{i+1} \vDash \mathfrak{M}_{i+1}$.

For the first inequality the argument is analogous. By Theorem 2.5, there exists a w-AD family $\mathcal{F}$ for F. As before, suppose for sake of contradiction that there is a space-aware resolution refutation π of F such that $\mathrm{CSp}(\pi) \leqslant w - k$. Let $\pi = (\mathfrak{M}_0, \ldots, \mathfrak{M}_\ell)$. We show, by induction on i, that for every $\mathfrak{M}_i$ there exists some $\alpha_i \in \mathcal{F}$ such that $\alpha_i \vDash \mathfrak{M}_i$, that is for each clause $C \in \mathfrak{M}_i$, $C\!\restriction_{\alpha_i} = 1$. The only difference with the previous argument is how to deal with the case when $\mathfrak{M}_{i+1} = \mathfrak{M}_i \cup \{C\}$ with $C \in F$. Let V be the set of variables in C not in $\mathrm{dom}(\alpha_i)$. By assumption $|V| \leqslant k$ and, as before, there exists $\beta \subseteq \alpha_i$ such that $|\mathrm{dom}(\beta)| \leqslant |\mathfrak{M}_i| \leqslant w - k$ and $\beta \vDash \mathfrak{M}_i$. Then, by the extension property of $\mathcal{F}$, we can extend β $|V|$-times to get a $\beta' \in \mathcal{F}$ that sets all the variables in C. Then by the consistency property of $\mathcal{F}$, $C\!\restriction_{\beta'} = 1$ and as before we can set $\alpha_{i+1} = \beta'$. It is straightforward to check that $\alpha_{i+1} \vDash \mathfrak{M}_{i+1}$. $\qquad\square$

3.2.1 Semantic Clause Space

Semantic resolution refutations are not really more efficient than usual resolution refutations from the point of view of the clause space, as the next theorem shows.

Theorem 3.3 ([ABRW02]). *Let F be an unsatisfiable CNF formula in n variables and suppose that every resolution refutation of F requires clause space $\geqslant c$. Then for every semantic space-aware resolution refutation π of F*

$$\mathrm{CSp}(\pi) \geqslant \frac{c}{2} . \qquad \square \tag{3.11}$$

Regarding the lower bounds we recall that Theorem 3.2 trivially generalizes to semantic clause space refutations and a similar proof can actually show lower bounds on the clause space of semantic resolution refutations of semiwide formulas.

Theorem 3.4 ([AD08, BK14]). *Let F be an unsatisfiable k-CNF formula and suppose that every resolution refutation of F requires width $\geqslant w$ and asymmetric width $\geqslant w'$. Then for every semantic space-aware resolution refutation π of F*

$$\mathrm{CSp}(\pi) > w - k , \tag{3.12}$$
$$\mathrm{CSp}(\pi) > w' . \tag{3.13}$$

Proof. The proof is exactly the same as that of Theorem 3.2 since in that proof we never used the actual inference rule of resolution but only its soundness. $\qquad\square$

Regarding the lower bounds for semantic resolution we have the following result.

Theorem 3.5 ([ABRW02]). *Let F be an unsatisfiable r-semiwide CNF formula. Then for every semantic space-aware resolution refutation π of F*

$$\mathrm{CSp}(\pi) \geqslant r . \qquad \square \tag{3.14}$$

3.3 Total Space

From the upper bound on clause space we immediately have the following upper bound on total space.

Corollary 3.1. *For every unsatisfiable CNF formula F in n variables there exists π a space-aware refutation of F such that*

$$\text{TSp}(\pi) \leqslant n^2 + n. \tag{3.15}$$

Proof. It follows immediately from the definition of total space, the fact that each clause can contain at most n literals and Theorem 3.10. □

Regarding total space lower bounds we have a result similar to the clause space-width inequality, see Theorem 3.2. The proof of the total space-width inequality is a bit more tricky and there are some good reasons for it to be more involved. We postpone such discussions for a while; let's see the result and its proof first.

Theorem 3.6 ([Bon16]). *Let F be an unsatisfiable k-CNF formula and suppose that every resolution refutation of F requires width $\geqslant w$ and asymmetric width $\geqslant w'$. Then for every resolution refutation π of F*

$$\text{TSp}(\pi) \geqslant \lfloor (w - k - 4)/4 \rfloor^2, \tag{3.16}$$

$$\text{TSp}(\pi) \geqslant \lfloor (w' - 2)/2 \rfloor^2. \tag{3.17}$$

Proof. Thanks to Lemma 2.2 it is enough to just prove the second inequality. Indeed if every resolution refutation of F requires width $\geqslant w$ then for every resolution refutation π of F,

$$2\text{aW}(\pi) + k \geqslant \text{aW}(\pi) + \max\{\text{aW}(\pi), k\} \geqslant w, \tag{3.18}$$

hence $\text{aW}(\pi) \geqslant \frac{1}{2}(w - k)$ and taking $w' = \frac{1}{2}(w - k)$ we immediately get the first inequality from the second.

By Theorem 2.6 there exists a non-empty $(w' - 1)$-BK family $\mathcal{F}$ for F. We show then that for every resolution refutation $\pi = (\mathfrak{M}_1, \ldots, \mathfrak{M}_\ell)$ there exists an index i such that $\mathfrak{M}_i$ contains at least $\lfloor (w' - 2)/2 \rfloor$ clauses each with at least $\lfloor (w' - 2)/2 \rfloor$ literals. This will immediately imply the second inequality of the theorem. Consider the set

$$S = \{C \text{ clause} : \exists \alpha \in \mathcal{F} \ C{\restriction}_\alpha = 0\}. \tag{3.19}$$

Since $\mathcal{F}$ is non-empty, then $\bot \in S$ and, by the consistency property of $\mathcal{F}$, no clause from F is in S. Hence, the set

$$A = \{i \in [\ell] : \exists C \in \mathfrak{M}_i \cap S \ |C| < \lfloor (w' - 2)/2 \rfloor\} \tag{3.20}$$

is non-empty. Let $t = \min A$ and let $C \in \mathfrak{M}_t \cap S$ be a clause of width strictly less than $\lfloor (w' - 2)/2 \rfloor$. Let $\alpha \in \mathcal{F}$ be a Boolean assignment that falsifies C and let α_C

be the Boolean assignment contained in α falsifying C and with domain assigning only the variables in C. Our goal is to show that there is some $i < t$ such that $|\mathfrak{M}_i \cap S| \geq \lfloor (w'-2)/2 \rfloor$. Since for every $i < t$ every clause in $\mathfrak{M}_i \cap S$ has width at least $\lfloor (w'-2)/2 \rfloor$, this will give the desired result.

Suppose then, for sake of contradiction, that for each $i < t$,

$$|\mathfrak{M}_i \cap S| < \lfloor (w'-2)/2 \rfloor . \tag{3.21}$$

We inductively find Boolean assignments $\alpha_0, \ldots, \alpha_t$ in $\mathcal{F}$ such that for each $i \leq t$, $\alpha_C \subseteq \alpha_i$ and $\alpha_i \vDash \mathfrak{M}_i \cap S$. This immediately gives a contradiction when we reach $i = t$, since α_C falsifies the clause $C \in \mathfrak{M}_t \cap S$ and $\alpha_t \supseteq \alpha_C$.

The base case $i = 0$ is trivial: $\mathfrak{M}_0 = \emptyset$, so we can put $\alpha_0 = \alpha_C$. For the inductive step from $\mathfrak{M}_i$ to $\mathfrak{M}_{i+1}$ we might have an axiom download, an erasure or an inference. In the cases of an erasure or axiom download it holds that $\mathfrak{M}_{i+1} \cap S \subseteq \mathfrak{M}_i \cap S$, since clauses from F are not in S. So in those cases we can just set $\alpha_{i+1} = \alpha_i$.

In the case of an inference step, suppose that $\mathfrak{M}_{i+1} = \mathfrak{M}_i \cup \{D \vee E\}$ where $D \vee E$ is the result of the resolution rule applied on the clauses $D \vee x$ and $E \vee \neg x$ in $\mathfrak{M}_i$ resolved on the variable x. If $D \vee E \notin S$ we have nothing to do, just set $\alpha_{i+1} = \alpha_i$. Otherwise suppose $D \vee E \in S$. With an argument totally analogous to Lemma 3.1 it is immediate to see that there exists a Boolean assignment β such that $\alpha_C \subseteq \beta \subseteq \alpha_i$, $\beta \vDash \mathfrak{M}_i \cap S$ and

$$|\text{dom}(\beta)| \leq |\text{dom}(\alpha_C)| + |\mathfrak{M}_i \cap S| < \lfloor (w'-2)/2 \rfloor + \lfloor (w'-2)/2 \rfloor = w'-2 . \tag{3.22}$$

Observe that for every Boolean assignment $\gamma \in \mathcal{F}$ such that $\beta \subseteq \gamma$ and every clause $C' \in \mathfrak{M}_i$ such that $\text{var}(C') \subseteq \text{dom}(\gamma)$, it holds that $C'{\upharpoonright}_\gamma = 1$. This is due to the fact that since $\text{var}(C') \subseteq \text{dom}(\gamma)$ then γ sets C' either to 0 or to 1. It cannot set it to 0 because otherwise $C' \in \mathfrak{M}_i \cap S$ and then $C'{\upharpoonright}_\beta = 1$; which is not possible since $\beta \subseteq \gamma$.

If there exists some variable y in $D \vee E$ not in $\text{dom}(\alpha_i)$ then, by the extension property of $\mathcal{F}$, we can extend β to $\beta_0, \beta_1 \in \mathcal{F}$ both extending β and respectively setting y to 0 and 1. If the literal y is in $D \vee E$ we set $\alpha_{i+1} = \beta_1$; if the literal $\neg y$ is in $D \vee E$ we set $\alpha_{i+1} = \beta_0$. It is straightforward to check that $\alpha_{i+1} \vDash \mathfrak{M}_{i+1} \cap S$.

Suppose then that $\text{var}(D \vee E) \subseteq \text{dom}(\alpha_i)$. If $x \in \text{dom}(\alpha_i)$ then by what we observed before we must have that $D \vee x{\upharpoonright}_{\alpha_i} = 1$ and $E \vee \neg x{\upharpoonright}_{\alpha_i} = 1$, hence, by the soundness of resolution $D \vee E{\upharpoonright}_{\alpha_i} = 1$. In this case we can just set $\alpha_{i+1} = \alpha_i$.

If $x \notin \text{dom}(\alpha_i)$ then, by the extension property of $\mathcal{F}$, we can extend β to $\beta_0 \in \mathcal{F}$ such that $\beta_0(x) = 0$. If $\text{var}(D \vee x) \subseteq \text{dom}(\beta_0)$ then, by what we observed before, it must be that $D \vee x{\upharpoonright}_{\beta_0} = 1$. Since $\beta_0(x) = 0$ then $D{\upharpoonright}_{\beta_0} = 1$ and hence we can just set $\alpha_{i+1} = \beta_0$. If $\text{var}(D \vee x) \not\subseteq \text{dom}(\beta_0)$ then there exists some variable z in $D \vee x$ not in $\text{dom}(\beta_0)$. Let $\beta_0' = \beta \cup \{x = 0\}$. We have that $|\text{dom}(\beta_0')| = |\text{dom}(\beta)| + 1 < w' - 1$ hence we can extend β_0' to some $\gamma_0, \gamma_1 \in \mathcal{F}$ such that γ_0 sets z to 0 and γ_1 sets z to 1. If the literal z is in D we set $\alpha_{i+1} = \gamma_1$; if the literal $\neg z$ is in D we set $\alpha_{i+1} = \gamma_0$. $\qquad \square$

We want to stress the fact that this proof is intrinsically different from the proof of Theorem 3.2. In particular in both proofs we have an axiom download case and an inference case. In this proof the easy case is the axiom download, while in the proof of Theorem 3.2 the easy case is the inference. If we, intuitively, believe that total space and *semantic* total space in resolution are separated then the fact that the inference case in the above proof is the '*hard*' case is to be expected. Otherwise (informally) the proof would have been valid for semantic total space too, while we expect this not to be the case. Indeed the question of equivalence between resolution and semantic resolution from the point of view of the total space was asked already in [ABRW02, Open Question 4].

There are unsatisfiable 3-CNF formulas, for instance random 3-CNF formulas F_n in n variables with a linear number of clauses (see Sect. 7.2) such that for each resolution refutation π of F,

$$W(\pi) = \Omega(n) \,, \tag{3.23}$$

and hence by the previous theorem requiring $\Omega(n^2)$ total space to be refuted by resolution. On the other hand, the trivial semantic resolution refutation π' of F is such that

$$\mathrm{TSp}(\pi') \leqslant 3|F| = \Omega(n) \,. \tag{3.24}$$

3.3.1 Semantic Total Space

Regarding the semantic total space we show two results; the first one tells us a way to prove good total space lower bounds for semantic resolution refutations and the second will be a generalization of Theorem 3.6 to a bounded version of semantic resolution.

Theorem 3.7 ([BGT16]). *Let F be an unsatisfiable r-semiwide CNF formula. Then, for every semantic space-aware resolution refutation $\pi = (\mathfrak{M}_0, \ldots, \mathfrak{M}_\ell)$ of F*

$$\mathrm{TSp}(\pi) \geqslant \lfloor r/2 \rfloor^2 \,. \tag{3.25}$$

More precisely, there exists an $\mathfrak{M}_i$ that contains at least $\lfloor r/2 \rfloor$ clauses each having at least $\lfloor r/2 \rfloor$ literals.

Proof. Let $F = F' \wedge W$ as in Definition 3.1 and let $\pi = (\mathfrak{M}_0, \ldots, \mathfrak{M}_\ell)$ be a semantic space-aware resolution refutation of F. Let $\mathfrak{M}_i^*$ be the set of clauses $C \in \mathfrak{M}_i$ such that $F' \nvDash C$, that is such that there exists a Boolean assignment satisfying F' and falsifying C. Take the first t such that there exists a clause $C \in \mathfrak{M}_i^*$ of width strictly less than $\lfloor r/2 \rfloor$. Fix such a clause C and let α be the Boolean assignment falsifying C with domain var(C). Then α is F'-consistent and $|\mathrm{dom}(\alpha)| = |C| < \lfloor r/2 \rfloor$.

As in Theorem 3.6, it is then enough to show that $|\mathfrak{M}_i^*| \geqslant \lfloor r/2 \rfloor$ for some $i < t$, since for every $i < t$ every clause in $\mathfrak{M}_i^*$ has width at least $\lfloor r/2 \rfloor$. For sake of contradiction suppose that $|\mathfrak{M}_i^*| < \lfloor r/2 \rfloor$ for all $i < t$. By induction we show that for each $i = 1, \ldots, t$ there exists some F'-consistent $\alpha_i \supseteq \alpha$ such that $\alpha_i \vDash \mathfrak{M}_i^*$. For $i = t$ this is the contradiction sought.

The base case $i = 0$ is trivial, just set $\alpha_0 = \alpha$. For the inductive step we distinguish between two possibilities: that is whether $\mathfrak{M}_{i+1}$ is obtained from $\mathfrak{M}_i$ by performing a semantic inference or an axiom download. For the semantic inference case, that is $\mathfrak{M}_i \vDash \mathfrak{M}_{i+1}$, we let α_{i+1} be any extension of α_i that satisfies F'. Then from the fact that $\alpha_{i+1} \vDash \mathfrak{M}_i^* \wedge F'$ it follows that $\alpha_{i+1} \vDash \mathfrak{M}_i$ and hence $\alpha_{i+1} \vDash \mathfrak{M}_{i+1}$. For the axiom download case, suppose that $\mathfrak{M}_{i+1} = \mathfrak{M}_i \cup \{D\}$ with D a clause from W. With an argument analogous to the one proving Lemma 3.1, we may assume without loss of generality that

$$|\mathrm{dom}(\alpha_i)| \leqslant |\mathrm{dom}(\alpha)| + |\mathfrak{M}_i^*| < r . \tag{3.26}$$

Hence by the r-semiwideness of F there is a F'-consistent $\alpha_{i+1} \supseteq \alpha_i$ such that $\alpha_{i+1} \vDash D$. Then clearly $\alpha_{i+1} \vDash \mathfrak{M}_{i+1}$.   □

We said that the total space-width inequality cannot hold for semantic resolution. This is correct but this inequality can be generalized easily to d-semantic resolution as follows.

Theorem 3.8 ([Bon15, BGT16]). *Let F be an unsatisfiable k-CNF formula and suppose that for every resolution refutation π of F, $\mathrm{W}(\pi) \geqslant w$ and $\mathrm{aW}(\pi) \geqslant w'$. Then for every d-semantic resolution refutation π of F*

$$\mathrm{TSp}(\pi) \geqslant \lfloor (w - k - 2d)/4 \rfloor^2 , \tag{3.27}$$

$$\mathrm{TSp}(\pi) \geqslant \lfloor (w' - d)/2 \rfloor^2 . \tag{3.28}$$

Proof. The proof is the same as for Theorem 3.6, except that we have to argue differently for the inference step. By Theorem 2.6 there exists a non-empty $(w' - 1)$-BK family $\mathcal{F}$ for F. We show then that for every resolution refutation $\pi = (\mathfrak{M}_1, \ldots, \mathfrak{M}_\ell)$ there exists an index i such that $\mathfrak{M}_i$ contains at least $\lfloor (w' - d)/2 \rfloor$ clauses each with at least $\lfloor (w' - d)/2 \rfloor$ literals. This will immediately imply the theorem.

The argument goes as in the proof of Theorem 3.6: we just have to reason differently for the d-semantic inference. We follow the notations from the proof of Theorem 3.6 and we show here just how to adapt that argument. Suppose that in the inductive step we have a d-semantic inference: $\mathfrak{M}_{i+1} \subseteq \mathfrak{M}_i \cup \{E\}$ where E is implied by clauses $D_1, \ldots, D_d \in \mathfrak{M}_i$. We may assume that we have some $\alpha_i \in \mathcal{F}$ such that $\alpha_i \vDash \mathfrak{M}_i \cap S$ and let $\beta \subseteq \alpha_i$ be of minimal size such that $\beta \vDash \mathfrak{M}_i \cap S$ and $\beta \supseteq \alpha_C$. Since $|\mathrm{dom}(\alpha_C)| < \lfloor (w' - d)/2 \rfloor$ and $|\mathfrak{M}_i \cap S| < \lfloor (w' - d)/2 \rfloor$, then

$$|\mathrm{dom}(\beta)| \leqslant |\mathrm{dom}(\alpha_C)| + |\mathfrak{M}_i \cap S| < w' - d . \tag{3.29}$$

Either D_1 is satisfied by α_i or it is not. If it is, let $\gamma_1 = \alpha_i$. If not, then D_1 cannot be in S, since α_i satisfies all members of $\mathfrak{M}_i \cap S$. It follows that D_1 is not falsified by α_i either, otherwise D_1 would be in S. Then, by the inductive hypothesis, D_1 will be satisfied by α_i. So D_1 thus must contain some literal not set by α_i. In this case let $\gamma_1 \in \mathcal{F}$ be an extension of β which satisfies this literal.

We have found $\gamma_1 \in \mathcal{F}$ which satisfies D_1 with $\beta \subseteq \gamma_1$. We then take a minimal partial assignment γ_1' contained in γ_1 such that $\gamma_1' \vDash D_1$ and $\gamma' \supseteq \beta$. We have that $|\mathrm{dom}(\gamma_1')| \leqslant |\mathrm{dom}(\beta)| + 1 < w' - d + 1$, so we can repeat the previous reasoning on γ_1' and D_2 instead of β and D_1, and again up to D_d. In this way we build a sequence of extensions $\gamma_1 \subseteq \gamma_2 \subseteq \cdots \subseteq \gamma_d$ in $\mathcal{F}$, finishing with γ_d, which satisfies each of $D_1, \ldots, D_d$ and thus also satisfies the inferred clause E. We then take $\alpha_{i+1} = \gamma_d$. $\square$

3.4 Space and Depth

It turns out that total space in resolution (and other reasonable proof systems) is polynomially equivalent to the *depth* of proofs [Raz16b]. The *depth* of a resolution refutation DAG π is just the longest path from $\bot$ to some input clause. Let us denote the depth of π by $D(\pi)$. This complexity measure has a characterization as families of assignments similar to the ones we saw for the width and the asymmetric width, see [Urq11a].

For the rest of this section consider a fixed unsatisfiable CNF formula F_n in n variables such that every resolution refutation of it requires depth $\geqslant D$. Using the previous characterization of depth, [Urq11a, Theorem 6.1] shows that there exists a resolution refutation π of F_n such that

$$\mathrm{VSp}(\pi) \leqslant D. \tag{3.30}$$

Now, [Urq11a, Problem 7.2] already asked in an informal way "how much" this inequality is not tight. Recently [Raz16b] showed that there is a quadratic gap between the two complexity measures. However it still might be possible that the two measures are polynomially related [Raz16b]. This is related to an open question about the relation between clause space and total space, see Question 3.5.

Regarding total space, as [Raz16b] observes, a simple modification of the argument for the clause space upper bound (Theorem 3.10) gives the following upper bound on clause space and total space in resolution.

Theorem 3.9 ([Raz16b]). *For every unsatisfiable CNF formula F in n variables such that every resolution refutation of it requires depth $\geqslant D$, there exists π a space-aware refutation of F such that*

$$\mathrm{CSp}(\pi) \leqslant D+1, \tag{3.31}$$
$$\mathrm{TSp}(\pi) \leqslant D(D+1). \quad \square \tag{3.32}$$

Now, clearly the width needed to refute a CNF formula F is always smaller than the depth needed to refute F. Then immediately from Theorem 3.6 and its applications, we have that eq. (3.31) is asymptotically tight. Moreover it is tight for many CNF formulas, e.g., random k-CNF formulas (see Sect. 7.2) and more generally for CNF formulas that have asymptotically optimal width lower bounds. Interestingly the depth also lower bounds the total space in resolution.

Theorem 3.10 ([Raz16b]). *For every unsatisfiable CNF formula F in n variables such that every resolution refutation of it requires depth $\geqslant D$, for every space-aware refutation π of F*

$$\mathrm{TSp}(\pi) \geqslant \widetilde{\Omega}(\sqrt{D}). \quad \square \tag{3.33}$$

3.5 Open Problems

Question 3.1. Are there any non-trivial total space lower bounds for stronger proof systems such as bounded-depth Frege, polynomial calculus or cutting planes?

We recall that for unrestricted Frege systems there is a linear upper bound (in the size of the CNF formula being refuted) on total space, see [ABRW02]. Regarding cutting planes some preliminary results on space are shown in [GPT15].

Question 3.2. Is there a family of k-CNF formulas in n variables and $n^{O(1)}$ clauses that have resolution refutations of size $n^{O(1)}$ but still such that for each resolution refutation π, $\mathrm{TSp}(\pi) = \Omega\left(n^2\right)$ or at least $\mathrm{TSp}(\pi) = \omega(n)$?

In [BW01] the authors showed that if a k-CNF formula in n variables has a resolution refutation of size S then it also has a refutation in which every clause has width at most $O(\sqrt{n \log S})$. In [BG01] it is shown that this result is almost optimal. That is, there are formulas in n variables, with polynomial-size resolution refutations but needing width $\Omega(\sqrt{n})$ to be refuted. Those formulas are the ordering principles and this width upper bound cannot be used together with Theorem 3.6 to prove an $\omega(n)$ total space lower bound. However, it is possible that for some minor modification of those formulas we still have polynomial-size resolution refutations but now we can prove a width lower bound that is $\omega(\sqrt{n})$. This is enough, together with Theorem 3.6, to answer the previous question. However the total space-width inequality will never show a total space lower bound of the form $\Omega\left(n^2\right)$ for such formulas.

Question 3.3. Is eq. (3.33) tight? That is, is there a resolution formula F_n in n variables such that every resolution refutation of it requires depth D and there is a resolution refutation π of F_n such that $\mathrm{TSp}(\pi) = \widetilde{\Theta}(\sqrt{D})$?

This very same question was asked already in [Raz16b] and indeed it seems not clear at all what should be the answer to it.

Question 3.4. Let F_n be an unsatisfiable CNF formula in n variables and let V be the minimal variable space required to refute it. Is there a resolution refutation π of F_n and a constant c such that

$$\mathrm{D}(\pi) \leqslant (V \cdot \log n)^c \; ? \tag{3.34}$$

This exact same question was asked in [Raz16b] and in a more informal way in [Urq11a].

Question 3.5. Let F_n be an unsatisfiable CNF formula in n variables and let V be the minimal variable space required to refute it. Is there a resolution refutation π of F_n and a constant c such that

$$\mathrm{CSp}(\pi) \leqslant (V \cdot \log n)^c \; ? \tag{3.35}$$

If we exchange the roles of clause space and variable space in the previous question then we know that it is false. Indeed [BNT13] showed that there are unsatisfiable 3-CNF formulas in n variables that can be refuted in clause space 2 but every resolution refutation of them requires variable space at least $\Omega\left(n/\log n\right)$. Now, as observed in [Raz16b], eq. (3.35) is clearly implied by eq. (3.34) but of course eq. (3.35) might actually be easier to prove.

History

The total space-width inequality in this form (Theorem 3.6) was proved in [Bon15, Bon16] although the proof idea essentially dates back to [BGT14]. The main difference between the two approaches is that in [BGT14] we were using a different family of assignments, the r-BGT families, somehow inspired by the (r, I)-BG families (Definition 4.5) we will see in the next chapter. The r-BGT families and the r-BK families although independently introduced shared a lot of properties and this allowed us to adapt the total space lower bounds proved in [BGT14] to the r-BK families.

Theorem 3.7, the total space lower bounds for semiwide formulas in semantic resolution, was proved in [BGT14] as a straightforward generalization of the total space lower bounds by [ABRW02] for two particular n-semiwide formulas: PHP_n^{n+1} and CT_n, the *only* total space lower bound known before [BGT14].

Chapter 4
Space in Polynomial Calculus

In this chapter we consider in detail the propositional proof system *polynomial calculus*, briefly introduced in Sect. 1.1.2. We introduce some complexity measures— size, degree, monomial space and total space—and we show some techniques useful to prove lower bounds on those.

The main result of this chapter is a theorem reducing the problem of proving monomial space lower bounds in polynomial calculus to the, supposedly, easier task of constructing some family of Boolean assignments with certain combinatorial properties (Theorem 4.2). Such families essentially play the same role as the w-AD families or w-BK families we saw in Sect. 2.3.1.

Recall that given a field $\mathbb{F}$, a set of variables V and a set of polynomials P in the ring $\mathbb{F}[V]$, the *polynomial calculus* is a way to certify that P does not have common zeros in $\mathbb{F}$. This is done by showing that $\langle P \rangle$, the ideal generated by P, is the whole ring $\mathbb{F}[V]$, or in other words that $1 \in \langle P \rangle$.

A *polynomial calculus refutation of P over $\mathbb{F}[V]$* is a sequence of polynomials $\pi = (p_1, \ldots, p_\ell)$ such that $p_\ell = 1$ and for each $p_i \in \pi$ either $p_i \in P$ or it is obtained by applying one of the following inference rules with premises in $\{p_1, \ldots, p_{i-1}\}$:

$$\frac{p \qquad q}{\alpha p + \beta q} \; \alpha, \beta \in \mathbb{F}, \qquad\qquad \frac{p}{xp} x \in V . \qquad (4.1)$$

Given a polynomial calculus refutation π, its *size* $\mathrm{S}(\pi)$ is the number of monomials (counted with repetitions) appearing in π.

Clearly if there is a polynomial calculus refutation of a set of polynomials P then $1 \in \langle P \rangle$ and hence P cannot have common zeros. The converse is in general not true. For instance consider the field of the real numbers $\mathbb{R}$ and $x^2 + 1 \in \mathbb{R}[x]$. It is well known that this polynomial does not have zeros in $\mathbb{R}$ and at the same time $1 \notin \langle x^2 + 1 \rangle$.

© Springer International Publishing AG, part of Springer Nature 2017
I. Bonacina, *Space in Weak Propositional Proof Systems*,
https://doi.org/10.1007/978-3-319-73453-8_4

4.1 From CNF Formulas to Polynomials

We mostly consider particular sets of polynomials P encoding CNF formulas, and for those P it will hold that P has a polynomial calculus refutation over $\mathbb{F}$ if and only if P has no common zeros. And this will hold regardless of the ground field $\mathbb{F}$.

Given a CNF formula F over a set of Boolean variables X and a field $\mathbb{F}$ we want to construct a set of variables V and a set of polynomials P in $\mathbb{F}[V]$ such that F is unsatisfiable if and only if P has no common zeros in $\mathbb{F}$.

Let's try first to take $V = X$. Since the variables in X can only have values in $\{0,1\}$ it is natural to require P to have zeros inside $\{0,1\}^X$. That is P will contain the set of polynomials $\{x^2 - x : x \in X\}$. These polynomials are usually called the *Boolean axioms* in $\mathbb{F}[X]$.

Then any polynomial calculus refutation ends with the constant polynomial 1, so it is somehow natural to think about it as the "trivially false" polynomial. Then the constant polynomial 0 will be the "trivially true" polynomial. So the meaning of $0, 1$ in this polynomial setting is the opposite of their meaning in the context of propositional Boolean formulas. With this in mind it is then natural to encode clauses $C = \bigvee_{x \in I} x \vee \bigvee_{y \in J} \neg y$, with $I, J \subseteq X$ as

$$p'_C = \prod_{x \in I}(1 - x) \cdot \prod_{y \in J} y . \tag{4.2}$$

With this choice, for every Boolean assignment α of the variables X, $C\restriction\alpha = 1$, that is α sets C to true, if and only if $p'_C\restriction\alpha = p'_C(\alpha) = 0$, that is α sets p'_C to true.

It is then natural to encode a CNF formula F as the set of polynomials

$$P'_F = \{p'_C : C \in F\} \cup \{x^2 - x : x \in X\} . \tag{4.3}$$

Then F is unsatisfiable if and only if P_F has no common zeros in $\mathbb{F}$.

There is only one issue with this approach: each of the polynomials in p'_C has to be written as a sum of monomials and by eq. (4.2) this sum may contain exponentially many polynomials. Indeed, if F is a k-CNF formula then each p'_C associated with the clauses of F contains at most 2^k monomials. If F has n variables and it has some clause with $\omega(\log n)$ positive literals then P'_F has super-polynomially many monomials. We need a more efficient encoding.

Take $V = X \cup \overline{X}$, where $\overline{X} = \{\bar{x} : x \in X\}$ where the $\bar{x}$ are new formal variables.[1] The semantic meaning of $\bar{x}$ is $1 - x$. Now we can redo most of the work we did before. The set of *Boolean axioms* in $\mathbb{F}[X \cup \overline{X}]$ is $B = \{x^2 - x, \bar{x}^2 - \bar{x}, x + \bar{x} - 1 : x \in X\}$. The encoding of eq. (4.2) becomes

$$m_C = \prod_{x \in I} \bar{x} \cdot \prod_{y \in J} y , \tag{4.4}$$

which now is a monomial. The CNF formula F is encoded as the set of polynomials

---

[1] This convenient way of translating CNF formulas into polynomials was introduced in [ABRW02].

$$P_F = \{m_C \ : \ C \in F\} \cup B,\qquad (4.5)$$

where B is the set of Boolean axioms for $\mathbb{F}[X \cup \overline{X}]$.

Most of the results we prove will hold for more general sets of polynomials of the form $P \cup B$ in $\mathbb{F}[X \cup \overline{X}]$ with no common zeros, that is we will not assume that P consists of monomials as we do for P_F.

Definition 4.1 (Boolean-Constrained). Given a set of variables V and a field $\mathbb{F}$, we call a set of polynomials P in $\mathbb{F}[V]$ *Boolean-constrained* if P contains the Boolean polynomials $\{x^2 - x \ : \ x \in V\}$.

Going back to the relation between a CNF formula F and the corresponding set of polynomials P_F in $\mathbb{F}[X \cup \overline{X}]$, we have that F is unsatisfiable if and only if P_F has no common zeros in $\mathbb{F}$. With a slight abuse of notation sometimes we say that π is a polynomial calculus refutation of F over $\mathbb{F}$ to mean that π is a polynomial calculus refutation of P_F in $\mathbb{F}[X \cup \overline{X}]$.[2] Now not only does a CNF formula F have a polynomial calculus refutation over $\mathbb{F}$ if and only if F is unsatisfiable, but we have the following stronger result.

Proposition 4.1. *Let $\mathbb{F}$ be a field, then the polynomial calculus over $\mathbb{F}$ p-simulates resolution, when CNF formulas F are encoded as polynomials using the encoding P_F in eq. (4.4).*

Proof. Polynomial calculus is a propositional proof system: its *soundness* follows from the fact that if P derives a polynomial q in polynomial calculus then $q \in \langle P \rangle$ and obviously q vanishes on the variety $V(P)$, that is the set of zeroes of P. The *completeness* of polynomial calculus follows since polynomial calculus simulates resolution.[3] Indeed it is easy to see how to simulate efficiently any instance of the resolution rule by some applications of the polynomial calculus rules, see Fig. 4.1. □

$$
\begin{array}{ccc}
\dfrac{\begin{array}{c} m_C \cdot \bar{x} \\ \vdots \\ m_C m_D \cdot \bar{x} \end{array} \quad \begin{array}{c} m_D \cdot x \\ \vdots \\ m_C m_D \cdot x \end{array}}{m_C m_D \cdot \bar{x} + m_C m_D \cdot x} & &
\dfrac{\begin{array}{c} x + \bar{x} - 1 \\ \vdots \end{array}}{m_C m_D x + m_C m_D \bar{x} - m_C m_D}
\end{array}
$$
$$\overline{\hspace{4cm} m_C m_D \hspace{4cm}}$$

Figure 4.1 Simulation of the rule $\dfrac{C \vee x, \quad D \vee \neg x}{C \vee D}$ in polynomial calculus

Before starting to investigate the proof complexity of polynomial calculus we describe a family of CNF formulas that we use later as a prototypical example.

---

[2] We will not consider results for the less efficient encoding P'_F although it has been studied in the proof complexity literature, see for instance [FLN+15].

[3] The completeness of polynomial calculus can be proved also as a corollary of Hilbert's Nullstellensatz [CLO97] or by the Gröbner basis algorithm [CEI96]. We do not require $\mathbb{F}$ to be algebraically closed due to the fact that we only consider sets of polynomials that are Boolean constrained.

4.1.1 Complete Tree Formulas CT_n

Let n be a natural number. The *complete tree formula* CT_n is the unsatisfiable CNF formula whose clauses are *all* the 2^n possible clauses with n distinct literals in the variables $X = \{x_1, \ldots, x_n\}$. For instance CT_2 is the following CNF formula:

$$\mathsf{CT}_2 = (x_1 \vee x_2) \wedge (\neg x_1 \vee x_2) \wedge (x_1 \vee \neg x_2) \wedge (\neg x_1 \vee \neg x_2) \qquad (4.6)$$

and its encoding as an unsatisfiable family of polynomials in $\mathbb{F}[X \cup \overline{X}]$ is

$$P_{\mathsf{CT}_2} = \{x_1 x_2,\ \bar{x}_1 x_2,\ x_1 \bar{x}_2,\ \bar{x}_1 \bar{x}_2,\ x_1^2 - x_1,\ x_2^2 - x_2,\ x_1 + \bar{x}_1 - 1,\ x_2 + \bar{x}_2 - 1\}\ . \quad (4.7)$$

Every (tree-like) resolution refutation π of CT_n has size $2^{n+1} - 1 = 2|\mathsf{CT}_n| - 1$. Moreover CT_n is n-semiwide, see Definition 3.1, and hence, for instance, every semantic resolution refutation π of it has clause space $\Omega(n)$, see Sect. 3.2.1, and total space $\Omega(n^2)$, see Sect. 3.3.1.

We use this CNF formula as a toy (but non-trivial) example for the complexity lower bounds and techniques we introduce in this chapter. More involved examples will be given in Part II.

4.2 Size and Degree

Given a field $\mathbb{F}$ and a set of variables V, consider an unsatisfiable set of polynomials P in $\mathbb{F}[V]$. Given a polynomial calculus refutation $\pi = (p_1, \ldots, p_\ell)$ over $\mathbb{F}$ of P, recall that its *size* $S(\pi)$ is the number of monomials (counted with repetitions) appearing in $p_1, \ldots, p_\ell$. The *degree* $\deg(\pi)$ is the maximum degree of a polynomial in π.

Given a CNF formula F in n variables, we saw in Sect. 2.2 that a width upper bound of w on the refutations of F implies a resolution size upper bound of $n^{O(w)}$, see Theorem 2.2. Similarly, given the polynomial encoding P_F of F, a degree upper bound of d on the polynomial calculus refutations of P_F implies a polynomial calculus size upper bound of $n^{O(d)}$, see [CEI96]. Both the upper bound in resolution and the one in polynomial calculus are tight, as shown by [ALN14].

In resolution we saw a size-width inequality, see Theorem 2.3. There is an analogous inequality between degree and size in polynomial calculus. Both results can be used to prove size lower bounds in the corresponding proof systems proving width or degree lower bounds.

Theorem 4.1 ([IPS99, Theorem 6.2][4]). *Let $\mathbb{F}$ be a field, V a set of n variables and P an unsatisfiable Boolean-constrained set of polynomials in $\mathbb{F}[V]$. Suppose that all the polynomials in P have degree at most k and for every polynomial calculus refutation π of P over $\mathbb{F}[V]$, $\deg(\pi) \geqslant d$. Then for every polynomial calculus refutation π*

---

[4] As observed in [MN15], the proof of this result holds also when $V = X \cup \overline{X}$ and $P \supseteq \{x + \bar{x} - 1\ :\ x \in X\}$.

of P over $\mathbb{F}[V]$

$$\log_2 S(\pi) = \Omega\left(\frac{(d-k)^2}{n}\right). \ \square \tag{4.8}$$

As for the width-size inequality for resolution, see eq. (2.10), this degree-size inequality is essentially optimal too [GL10b].

Proving degree lower bounds is usually more difficult than proving width lower bounds and, interestingly, much of this difficulty depends on the characteristic[5] char($\mathbb{F}$) of the ground field $\mathbb{F}$. If char($\mathbb{F}$) $\neq$ 2 then a Fourier-like transformation can be used to reduce degree lower bounds to width lower bounds [BI10]. A more general technique to prove degree lower bounds, working also if char($\mathbb{F}$) = 2, was introduced in [AR01] and generalized in [GL10a, MN15].

In this book we do not prove any polynomial calculus size or degree lower bounds; instead we focus on the space complexity of polynomial calculus and in particular on monomial space.

4.3 Space

In Chap. 3 we defined *space-aware* resolution derivations. For polynomial calculus we proceed in a similar way following [ET01, ABRW02]. Formally, given a field $\mathbb{F}$, a set of variables V and an unsatisfiable set of polynomials P in $\mathbb{F}[V]$, a *space-aware* polynomial calculus refutation of P over $\mathbb{F}$ is a sequence $\pi = (\mathfrak{M}_0, \ldots, \mathfrak{M}_\ell)$ of sets of polynomials in $\mathbb{F}[V]$, where $\mathfrak{M}_0$ is the empty set, $\mathfrak{M}_\ell$ contains the constant polynomial 1, and each $\mathfrak{M}_{i+1}$ is derived from $\mathfrak{M}_i$ in one of the following three ways:

- $\mathfrak{M}_{i+1} = \mathfrak{M}_i \cup \{p\}$, where p is a polynomial from F (**Axiom Download**);
- $\mathfrak{M}_{i+1} \subseteq \mathfrak{M}_i$ (**Erasure**);
- $\mathfrak{M}_{i+1} = \mathfrak{M}_i \cup \{p\}$ where p follows from some polynomials in $\mathfrak{M}_i$ by the polynomial calculus inference rules, see eq. (4.1) (**Inference**).

As for space-aware resolution, we call the $\mathfrak{M}_i$s *memory configurations*. It is immediate to see that, as for resolution, polynomial calculus and its space-aware version are p-equivalent.

Given a set of polynomials S, its *monomial* space MSp(S) is the number of <u>distinct</u> monomials occurring in S.[6] Given a sequence of sets $\pi = (\mathfrak{M}_0, \ldots, \mathfrak{M}_\ell)$, the monomial space of π is

$$\text{MSp}(\pi) = \max_{i\in[\ell]} \text{MSp}(\mathfrak{M}_i). \tag{4.9}$$

---

[5] Recall that the characteristic of a field $\mathbb{F}$ is the minimum positive number of times we need to add 1 (the multiplicative identity of $\mathbb{F}$) to get 0 (the additive identity of $\mathbb{F}$). If no number of 1s can sum up to 0 then $\mathbb{F}$ has characteristic 0. The characteristic of a field is always 0 or a prime number.

[6] A monomial is product of variables in $X \cup \overline{X}$.

Let $\mathbb{F}$ be a field and X a set of variables. A crucial feature of polynomial calculus refutations of a set of Boolean-constrained polynomials P in $\mathbb{F}[X \cup \overline{X}]$ such that $P \supseteq \{x + \bar{x} - 1 \ : \ x \in X\}$ is that sets of polynomials with large monomial space can be transformed into equivalent sets of polynomials with much smaller monomial space. For instance the set of polynomials $\{x_1, \ldots, x_m\}$ can be transformed into the equisatisfiable single polynomial $1 - \prod_{i \in [m]} \bar{x}_i$ just using intermediate memory configurations with $m + O(1)$ monomials. This observation—at the core of the following monomial space upper bound—suggests somehow that getting unconditional monomial space lower bounds is a much less trivial task than, for instance, getting unconditional clause space lower bounds in resolution. An easier task is to prove trade-offs, say between monomial space and size (or degree, for instance) and indeed in the literature there are many such results, see for example the survey [Nor15, Sect. 4.4].

Proposition 4.2 ([ABRW02, Theorem 4.2]). *Let $\mathbb{F}$ be a field and $X = \{x_1, \ldots, x_n\}$ be a set of variables. There exists a space-aware polynomial calculus refutation π of CT_n over $\mathbb{F}[X \cup \overline{X}]$ such that*

$$\mathrm{MSp}(\pi) \leqslant \frac{2}{3}n + 6 \,. \tag{4.10}$$

Hence for any unsatisfiable CNF formula F_n with variables in a set X of size n and every field $\mathbb{F}$, there exists a space-aware polynomial calculus refutation π of F_n over $\mathbb{F}[X \cup \overline{X}]$ such that

$$\mathrm{MSp}(\pi) \leqslant \frac{2}{3}n + 8 \,. \tag{4.11}$$

Proof (sketch). The first part of this statement is from [ABRW02]. For the second part it is enough to show that given a polynomial calculus refutation π of CT_n such that $\mathrm{MSp}(\pi) = s$ there is a polynomial calculus refutation π' of F with $\mathrm{MSp}(\pi') \leqslant s + 2$. This is due to the following fact. Every time there is an axiom download in π of a high-degree monomial $m \in P_{CT_n}$, there exists a monomial $m' \in P_F$ such that $m = m' \cdot m''$, where P_{CT_n} and P_F are the polynomial encodings of CT_n and F in $\mathbb{F}[X \cup \overline{X}]$ according to eq. (4.4). Then, using the polynomial calculus inference rules we can obtain m from m' by repeatedly multiplying by the variables in m'' one by one, and erasing the intermediate monomials. $\square$

Analogously to the total space in resolution, we have the notion of *total space* $\mathrm{TSp}(S)$ of a set of polynomials S. It is the total number of occurrences of variables in S. The total space $\mathrm{TSp}(\pi)$ of a sequence of sets of polynomials $\pi = (\mathfrak{M}_0, \ldots, \mathfrak{M}_\ell)$ is

$$\mathrm{TSp}(\pi) = \max_{i \in [\ell]} \mathrm{TSp}(\mathfrak{M}_i) \,. \tag{4.12}$$

Thanks to Proposition 4.2 we immediately have an upper bound on total space in polynomial calculus.

Proposition 4.3. *For any unsatisfiable CNF formula F_n with variables in X of size n and every field $\mathbb{F}$, there exists a space-aware polynomial calculus refutation π of the polynomial encoding of F_n over $\mathbb{F}[X \cup \overline{X}]$ such that*

$$\mathrm{TSp}(\pi) \leqslant \frac{2}{3}n^2 + 8n . \tag{4.13}$$

We will not cover total space lower bounds in polynomial calculus since at the moment of writing this book there are just two unconditional results known: a total space lower bound in polynomial calculus for CT_n and one for the *pigeonhole principle* PHP_n^m. Both results are proven in [ABRW02, Theorem 5.1]. The one for PHP_n^m is mentioned in more detail in Sect. 5.1. Here we recall the one for CT_n.

Proposition 4.4 ([ABRW02, Theorem 5.1]). *Let $\mathbb{F}$ be a field and $X = \{x_1, \dots, x_n\}$ a set of variables. Every polynomial calculus refutation[7] π of CT_n is such that*

$$\mathrm{TSp}(\pi) \geqslant \frac{3}{64}n^2 . \qquad \Box \tag{4.14}$$

As for the monomial space, there are some trade-offs known in the literature, see for example [BNT13].

4.4 Semantic Polynomial Calculus

In Sect. 3.1 we introduced the notion of semantic resolution refutations. Here we introduce an analogous concept for polynomial calculus: *I-semantic* polynomial calculus refutations, where I is an ideal.

Let X be a set of variables, $\mathbb{F}$ a field and S a set of polynomials in $\mathbb{F}[V]$. Recall that by $\langle S \rangle$ we denote the ideal generated by S in $\mathbb{F}[V]$ and given two ideals in $\mathbb{F}[V]$, I and J, their sum is $I + J = \{a + b \; : \; a \in I \text{ and } b \in J\}$. Using these concepts we can easily extend space-aware polynomial calculus refutations to what we call *I-semantic* (space-aware) polynomial calculus refutations.[8]

Definition 4.2 (*I-Semantic Refutation*). Given a field $\mathbb{F}$, a set of variables V, an unsatisfiable set of polynomials P and an ideal I in $\mathbb{F}[V]$. An *I-semantic* polynomial calculus refutation of P over $\mathbb{F}[V]$ is a sequence of sets of polynomials $\pi = (\mathfrak{M}_0, \dots, \mathfrak{M}_\ell)$ such that $\mathfrak{M}_0 = \emptyset$, $1 \in \mathfrak{M}_\ell$ and for all $i \leqslant \ell$,

$$\mathfrak{M}_i \subseteq \langle \mathfrak{M}_{i-1} \rangle + I + \langle p \rangle , \tag{4.15}$$

for some $p \in P$.[9]

---

[7] Eq. 4.14 actually holds for *I*-semantic polynomial calculus refutations π where I is the ideal generated by the Boolean axioms in $\mathbb{F}[X \cup \overline{X}]$, see Sect. 4.4

[8] This generalizes *semantical* polynomial calculus refutations from [ABRW02]: in our notation they are 0-semantic polynomial calculus refutations.

[9] We use the convention that $\langle \emptyset \rangle = 0$.

Intuitively this definition says that polynomials from I are always available in any step of the proof, so there is no need to store them in any $\mathfrak{M}_i$. That is we give away "for free" polynomials in I. Moreover in a single step one can reach "semantically" any polynomial in the ideal generated by the sum of the ideals in eq. (4.15) without storing the (possibly) complicated polynomials needed to obtain it. For instance given an unsatisfiable Boolean-constrained set of polynomials $P = \{p_1, \ldots, p_\ell\}$, then

$$(\emptyset, \{p_1\}, \{p_1, p_2\}, \ldots, \{p_1, \ldots, p_{\ell-1}\}, \{1\}) \tag{4.16}$$

is a perfectly valid 0-semantic polynomial calculus refutation of P. If P is not Boolean-constrained but just unsatisfiable the previous may or may not be a valid 0-semantic polynomial calculus refutation depending on the field $\mathbb{F}$.

When considering Boolean assignments in the context of I-semantic polynomial calculus refutations it is convenient to consider Boolean assignments respecting the ideal I.

Definition 4.3 ($\models_I$). Let $\mathbb{F}$ be a field and V be a set of variables. Given a family H of Boolean assignments over X, an ideal I and a set of polynomials S in $\mathbb{F}[V]$, we write $H \models_I S$ if for every Boolean assignment $\alpha \in H$, and every polynomial $p \in S$, $p\!\restriction_\alpha \in I$. If $H \models_I I$ we say that H is I-consistent.

4.5 Monomial Space Lower Bounds

Before introducing all the machinery to prove monomial space lower bounds, let's spend a few words on why some simpler naïve approach has no hope to work. A naïve approach could consist, for example, of trying to mimic the approach followed for the proof of Theorem 3.2. In this context, it will ultimately rely on the following property: given any polynomial p, if there is a Boolean assignment α that satisfies p, that is such that $p\!\restriction_\alpha = 0$, then there exists some Boolean assignment α' still satisfying p and such that $|\mathrm{dom}(\alpha')| \leqslant \mathrm{MSp}(p)$. Unfortunately this property is *false*: for instance $p = 1 - \prod_{i=1}^r x_i$ has just two monomials but any α zeroing it must have $|\mathrm{dom}(\alpha)| \geqslant r$. This phenomenon does not occour if we consider families of Boolean assignments, consisting of *many* Boolean assignments with a combinatorial structure we call *flippable products*. Then the proof of the monomial space lower bound we show, Theorem 4.2, and Theorem 3.2 have a similar structure, but instead of using Boolean assignments and the r-AD families we use *flippable products* and sets of such flippable products we call (r, I)-*BG families*.

Preliminary to the notion of (r, I)-BG families is the notion of *product-families*. We follow somewhat standard notations for Boolean assignments; anyway the reader might want to check their definitions in the Notation section on p. xv.

Definition 4.4 (Product-Families). Let V be a set of variables and let $H_1, \ldots, H_t$ be non-empty pairwise domain-disjoint[10] sets of Boolean assignments over V. The

---

[10] That is such that for each $i \neq j$, each $\alpha \in H_i$ and $\alpha' \in H_j$, $\mathrm{dom}(\alpha) \cap \mathrm{dom}(\alpha') = \emptyset$.

product-family $H = H_1 \otimes \cdots \otimes H_t$ is the following set of Boolean assignments over V:

$$H = H_1 \otimes \cdots \otimes H_t = \{\alpha_1 \cup \cdots \cup \alpha_t \ : \ \alpha_i \in H_i\}, \qquad (4.17)$$

or if $t = 0$ then H is a set containing just the empty Boolean assignment λ, that is $H = \{\lambda\}$. Moreover, we say that H is *flippable* if for every $i \in [t]$, $\alpha \in H_i$ and $x \in \text{dom}(\alpha)$ there exists $\alpha' \in H_i$ such that $\alpha'(x) = 0$.

We call the H_is *factors* of H and the *rank* of H, $\|H\|$, is the number of factors of H different from $\{\lambda\}$. We call a product-family whose factors are flippable a *flippable product-family* or simply a *flippable product*.

Notice that the same set of Boolean assignments can correspond to many product-families: in particular each family of Boolean assignments can be seen as a product of just one single factor. When we talk about a flippable product H we mean always that there is a particular fixed representation H as a product: say $H = H_1 \otimes \cdots \otimes H_t$. We do not count the $\{\lambda\}$ factors in the rank since they do not carry any additional information: the set of Boolean assignments corresponding to $H \otimes \{\lambda\}$ always coincides with H. Given two product-families H and H' we write $H' \sqsubseteq H$ if and only if each factor of H' different from $\{\lambda\}$ is also a factor of H. In particular $\{\lambda\} \sqsubseteq H$ for any product-family H.

Definition 4.5 ((r,I)-BG Families [BG15, Definition 3.4][11]). Let V be a set of variables, $\mathbb{F}$ a field, r an integer, P a set of polynomials and I an ideal in $\mathbb{F}[V]$. A family of flippable products $\mathcal{F}$ is an (r,I)-BG *family* for P if and only if for every $H \in \mathcal{F}$ the following conditions hold:

1. H is I-consistent, that is for every $p \in I$ and every $\alpha \in H$, $p\restriction_\alpha \in I$ (**Consistency**);
2. for each $H' \sqsubseteq H$, $H' \in \mathcal{F}$ (**Restriction**);
3. if $\|H\| < r$, then for each $p \in P$ there exists a flippable product $H' \in \mathcal{F}$ such that

   a. $H' \sqsupseteq H$, and
   b. $H' \vDash_I p$, that is for every $\alpha \in H'$, $p\restriction_\alpha \in I$ (**Extension**).

This definition shares some similarities with the w-AD and w-BK families we saw in Sect. 2.3.1 but, unlike those, the notion of (r,I)-BG family doesn't seem to characterize any complexity measure in polynomial calculus, e.g., the degree.

The main property of (r,I)-BG families is that they can be used to prove I-semantic polynomial calculus monomial space lower bounds.

Theorem 4.2 ([BG15, Theorem 3.5]). *Let V be a set of variables, $\mathbb{F}$ a field, $r \geqslant 1$ an integer, P an unsatisfiable set of polynomials and I a proper ideal in $\mathbb{F}[V]$. Suppose that there exists a non-empty (r,I)-BG family $\mathcal{F}$ for P. Then for every I-semantic polynomial calculus refutation π of P,*

$$\text{MSp}(\pi) \geqslant \lfloor r/4 \rfloor. \qquad (4.18)$$

---

[11] We follow the convention seen already in Sect. 2.3.1. That is we call families of Boolean assignments with particular combinatorial properties by the names of the authors that introduced them. In this case BG stands for Bonacina and Galesi in [BG13].

In this theorem we did not make any assumption on the structure of the set of initial polynomials P. If we have some additional assumptions on P it is possible to have an analogous result requiring the existence of a non-empty (r,I)-BG family just for a subset of P. In particular, this might be useful when we have some monomials of high degree in P.

Theorem 4.3. *Let V be a set of variables, $\mathbb{F}$ a field, $r \geqslant 1$ an integer, P an unsatisfiable set of polynomials and I a proper ideal in $\mathbb{F}[V]$. Let $P = P_1 \cup P_2$ and suppose that the following conditions hold:*

1. there exists a non-empty (r,I)-BG family $\mathcal{F}$ for P_1;
2. for every $H \in \mathcal{F}$ s.t. $\|H\| < r$ and every $p \in P_2$ either $H \vDash_I p$ or there exists $H' = H \otimes H_p \in \mathcal{F}$, with $\|H_p\| = 1$, and there exists $\alpha \in H_p$, $p\!\restriction_\alpha \in I$.

Then for every I-semantic polynomial calculus refutation π of P

$$\mathrm{MSp}(\pi) \geqslant \lfloor r/4 \rfloor \ . \tag{4.19}$$

The proofs of the previous two results are a bit technical and will be given in Sect. 4.5.2. It is better, before proving them, to familiarize ourselves a bit with the notion of (r,I)-BG families. First we prove that these families can be used also to prove total space lower bounds in resolution.[12] Secondly, we see how to apply the results to our "toy" example CT_n. More involved applications are shown in Chap. 5 and Chap. 7.

Proposition 4.5 ([BBG$^+$17]). *Let F be an unsatisfiable CNF formula in the variables X and let I be either the ideal 0 or the ideal generated by $\{x^2 - x : x \in X\}$ in $\mathbb{F}[X \cup \overline{X}]$. Suppose there exists a non-empty (r,I)-BG family for the polynomial encoding of F in $\mathbb{F}[X \cup \overline{X}]$, see eq. (4.5). Then for each resolution refutation π of F*

$$\mathrm{TSp}(\pi) \geqslant \lfloor r/2 - 1 \rfloor^2 \ . \tag{4.20}$$

Proof. To prove this theorem it is enough to show that given $\mathcal{F}$ a non-empty (r,I)-BG family for P_F we can construct an $(r-1)$-BK family of Boolean assignments $\mathcal{L}$ for F and then use Theorem 2.6 and Theorem 3.6.

Let $\mathcal{L}$ be the set of all the Boolean assignments α over X that appear in some flippable product H of $\mathcal{F}$ of rank at most $r-1$, that is

$$\mathcal{L} = \{\alpha \ : \ \exists H \in \mathcal{F} \ \exists \alpha' \in H \ \alpha \subseteq \alpha' \text{ and } \|H\| \leqslant r - 1\} \ .^{13} \tag{4.21}$$

This family $\mathcal{L}$ is non-empty (since for example the empty Boolean assignment $\lambda \in \mathcal{L}$) and, we claim, it is an $(r-1)$-BK family for F. That is we have to show it satisfies the *consistency* and *extension* properties of Definition 2.4. We use the notations from

---

[12] Thanks to Theorem 3.6 we know that total space lower bounds will follow from width lower bounds but that proof is somehow a bit abstract. The (r,I)-BG families are a more concrete witness of the fact that the total space is large.

[13] The reason we require $\alpha \subseteq \alpha'$ and not just $\alpha \in H$ is that the Boolean assignments in H are over a larger set of variables $X \cup \overline{X}$ and not just X.

Sect. 4.1; in particular given a clause C, m_C is its translation into a monomial in $\mathbb{F}[X \cup \overline{X}]$ and P_F is the translation of F as a set of polynomials in $\mathbb{F}[X \cup \overline{X}]$.

For the *consistency* property of $\mathcal{L}$ assume, for sake of contradiction, that there exists $\alpha \in \mathcal{L}$ such that α falsifies some clause C in F. That is $C\lceil_\alpha = 0$ and hence $m_C\lceil_\alpha = 1$. By definition of $\mathcal{L}$, there exists $H \in \mathcal{F}$ and $\alpha' \in H$ such that $\alpha \subseteq \alpha'$ and $\|H\| \leq r - 1$. By the extension property of $\mathcal{F}$, then there exists an $H' \sqsupseteq H$ such that $H' \models_I m_C$. In particular there exists some Boolean assignment $\alpha'' \supseteq \alpha' \supseteq \alpha$ such that $m_C\lceil_{\alpha''} \in I$. Hence $1 \in I$, which contradicts the fact that I is a proper ideal.

For the *extension* property of $\mathcal{L}$, consider $\alpha \in \mathcal{L}$, $\beta \subseteq \alpha$ with $|\mathrm{dom}(\beta)| < r - 1$ and x a variable from F not in $\mathrm{dom}(\alpha)$. Since $\beta \subseteq \alpha$ and $\alpha \in \mathcal{L}$ there must exist some $H \in \mathcal{L}$ and a $\beta' \in H$ such that $\beta \subseteq \beta'$ and $\|H\| \leq |\mathrm{dom}(\beta)| < r - 1$. By the extension property of $\mathcal{F}$, there exists some flippable product $H' \in \mathcal{F}$ such that $H' \sqsupseteq H$ and $H' \models_I x + \bar{x} - 1$. Since $x + \bar{x} - 1 \notin I$ we must have that every Boolean assignment in H' sets $x + \bar{x} - 1$ to 0, that is assigns x. Then, by definition of flippable product, all the factors of H' are domain-disjoint so there will be only one factor assigning x. By taking restrictions in $\mathcal{F}$, we can then suppose that $\|H'\| = \|H\| + 1 \leq r - 1$. All the Boolean assignments $\gamma \in H'$ extend β and all of them assign x. So it is enough to show that there is a $\gamma_0 \in H'$ setting $x = 0$ and another setting $x = 1$. By construction both γ_0 and γ_1 belong to $\mathcal{L}$, when restricted to X. If $\gamma_1 \in H'$ sets $x = 1$ then by the flippability condition on H' there exists $\gamma_0 \in H'$ setting $x = 0$. Suppose then there is $\gamma_0 \in H'$ that sets $x = 0$, then since $H' \models_I x + \bar{x} - 1$, it sets $\bar{x} = 1$. By the flippability condition on H' then there exists $\gamma_1 \in H'$ that sets $\bar{x} = 0$ and it also sets $x + \bar{x} - 1$ to 0, so γ_1 sets $x = 1$. $\square$

4.5.1 CT_n Requires Large Monomial Space

As a first example application of Theorem 4.2 and Theorem 4.3, we show how to prove a monomial space lower bound for CT_n.

Theorem 4.4 ([ABRW02, Corollary 4.19]). *Let $\mathbb{F}$ be a field, $X = \{x_1, \ldots, x_n\}$ and B the set of Boolean axioms in $\mathbb{F}[X \cup \overline{X}]$. Every $\langle B \rangle$-semantic polynomial calculus refutation π of CT_n over $\mathbb{F}[X \cup \overline{X}]$ is such that*

$$\mathrm{MSp}(\pi) \geq \lfloor n/4 \rfloor . \tag{4.22}$$

Proof. Use Theorem 4.3 with $P_1 = B$ and P_2 the degree n monomials in the polynomial encoding of CT_n in $\mathbb{F}[X \cup \overline{X}]$. The set P_1 is the set for which we have to build an $(n, \langle B \rangle)$-BG family $\mathcal{F}$. For each $i \in [n]$, let α_i and α'_i be the following two Boolean assignments of domain $\{x_i, \bar{x}_i\}$:

$$\alpha_i(x_i) = 1 , \quad \alpha_i(\bar{x}_i) = 0 , \tag{4.23}$$
$$\alpha'_i(x_i) = 0 , \quad \alpha'_i(\bar{x}_i) = 1 . \tag{4.24}$$

Let $H_i = \{\alpha_i, \alpha'_i\}$. By construction H_i is clearly flippable and $\langle B \rangle$-consistent. Let $\mathcal{F}$ be the family of all flippable products $H = \bigotimes_{i \in A} H_i$ for some set $A \subseteq [n]$. It is immediate to check that $\mathcal{F}$ is an $(n, \langle B \rangle)$-BG family. Regarding the second hypothesis of Theorem 4.3, let $H \in \mathcal{F}$ with $\|H\| < n$ and $m \in P_2$. We have that in m there is a variable, say x_i, not assigned by any Boolean assignment in H; then clearly there is a Boolean assignment in $H \otimes H_i$ satisfying m. It will be α_i if $\bar{x}_i$ appears in m or α'_i if x_i appears in m. □

4.5.2 Proofs of Theorem 4.2 and Theorem 4.3

Let's prove then Theorem 4.2 and Theorem 4.3. We need to introduce a new notion: the concept of 2-*merge*. On a very high level a 2-merge on a product-family H is a new product-family Z whose factors are obtained by '*merging*' disjoint pairs of factors from H.

Definition 4.6 (2-Merge). Let $H = H_1 \otimes \cdots \otimes H_t$ be a product-family. A 2-*merge* on H is a product-family $Z = Z_{J_1} \otimes \cdots \otimes Z_{J_r}$, where $J_1, \ldots, J_r$ are pairwise disjoint subsets of $[t]$ of size 1 or 2 such that for each $J_i = \{j, k\}$

$$Z_{J_i} = (A \otimes H_k) \cup (H_j \otimes B) , \tag{4.25}$$

where $A \subseteq H_j$ and $B \subseteq H_k$ are non-empty. (If $j = k$ then $Z_{J_i} = H_j$.)

Notice that, in the previous definition, if H is a flippable product then Z is also a flippable product. Moreover, given any ideal I, if H is I-consistent then Z is I-consistent.

To clarify the notion of 2-merge, let's see an example that will be useful both in the proof of Theorem 4.3 and in the proof of Lemma 4.1.

Example 4.1. Let m be a monomial and $H = H_1 \otimes H_2$ be a flippable product such that $\mathrm{vars}(m) \cap \mathrm{dom}(H_i) \neq \emptyset$ for $i = 1, 2$. Let $O_{m,i} = \{\alpha \in H_i : m \!\restriction_\alpha = 0\}$. Since H is flippable then $O_{m,i}$ is non-empty and then

$$Z = Z_{\{1,2\}} = (O_{m,1} \otimes H_2) \cup (H_1 \otimes O_{m,2}) \tag{4.26}$$

is a 2-merge on H. Moreover Z is a product-family since it has only one factor: $Z_{\{1,2\}}$.

As in [ABRW02] a key property in the proof of Theorem 4.2 is a "*Locality Lemma*".[14] Informally, this lemma asserts that if a set S of polynomials is satisfiable by a 2-merge on a product-family H, then it is possible to build a new 2-merge Z' on a new product-family H' such that Z' still satisfies S and $H' \sqsubseteq H$ has rank upper bounded by the number of distinct monomials in S.

---

[14] Lemma 4.1 is a generalization of analoguos results in [ABRW02, FLN+15, BG13]. The way we present it is based on [BG15].

Lemma 4.1 (Locality Lemma). *Let $\mathbb{F}$ be a field, V a set of variables, I an ideal and S a set of polynomials in $\mathbb{F}[V]$. Given a non-empty flippable product H and a 2-merge Z on H such that $Z \vDash_I S$. Then there exist a non-empty flippable product $H' \sqsubseteq H$ and a non-empty 2-merge Z' on H' such that $Z' \vDash_I S$ and $\|H'\| \leqslant 4 \cdot \mathrm{MSp}(S)$.*

The proof of this lemma is a bit technical so let's see first how to use it to prove Theorem 4.2 and Theorem 4.3, restated below for convenience.

Restated Theorem 4.2 ([BG15, Theorem 3.5]) *Let V be a set of variables, $\mathbb{F}$ a field, $r \geqslant 1$ an integer, P an unsatisfiable set of polynomials and I a proper ideal in $\mathbb{F}[V]$. Suppose that there exists a non-empty (r,I)-BG family $\mathcal{F}$ for P. Then for every I-semantic polynomial calculus refutation π of P,*

$$\mathrm{MSp}(\pi) \geqslant \lfloor r/4 \rfloor \ . \tag{4.18}$$

Proof. Let $\pi = (\mathfrak{M}_0, \dots, \mathfrak{M}_\ell)$ be an I-semantic polynomial calculus refutation of P and assume, for sake of contradiction, that $\mathrm{MSp}(\pi) < \lfloor r/4 \rfloor$. Moreover suppose that

for every $i = 0, \dots, \ell$, there exists a non-empty flippable product $H_i \in \mathcal{F}$ and a non-empty 2-merge Z_i on H_i such that $Z_i \vDash_I \mathfrak{M}_i$.

This claim immediately implies a contradiction: when $i = \ell$ it means that there exists some Boolean assignment $\alpha \in Z_\ell$ such that for every polynomial $p \in \mathfrak{M}_\ell$, $p\!\restriction_\alpha \in I$. However $1 \in \mathfrak{M}_\ell$, hence $1 \in I$, which is impossible since I is a proper ideal in $\mathbb{F}[V]$.

We prove the previous claim by induction on $i = 0, \dots, \ell$. For the base case $i = 0$, set $H_0 = \{\lambda\} \in \mathcal{F}$ and $Z_0 = H_0$. Then, trivially $\mathfrak{M}_0 = \emptyset$, so $Z_0 \vDash_I \mathfrak{M}_0$. For the inductive step, let $\mathfrak{M}_{i+1} \subseteq \langle \mathfrak{M}_i \rangle + I + \langle p \rangle$ with $p \in P$. By Lemma 4.1, used with parameters $H = H_i$, $Z = Z_i$ and $S = \mathfrak{M}_i$, we get a non-empty $H' \sqsubseteq H_i$ and a non-empty 2-merge Z' on H' such that $Z' \vDash_I \mathfrak{M}_i$ and $\|H'\| \leqslant 4\mathrm{MSp}(\mathfrak{M}_i)$. Then, by the restriction property of $\mathcal{F}$, $H' \in \mathcal{F}$. Moreover, we assumed that $\mathrm{MSp}(\mathfrak{M}_i) < \lfloor r/4 \rfloor$, so $\|H'\| \leqslant r - 4 < r$ and, by the extension property of $\mathcal{F}$ applied to H' and p, there exists $H_{i+1} \in \mathcal{F}$ such that $H_{i+1} \sqsupseteq H'$ and $H_{i+1} \vDash_I p$. Let $H_{i+1} = H' \otimes H_p$ where H_p is the flippable product collecting all the factors of H_{i+1} not in H'. Set $Z_{i+1} = Z' \otimes H_p$. It is a 2-merge on H_{i+1} due to the fact that Z' is a 2-merge on H' and by the definition of H_p. Finally, $H_{i+1} \vDash_I I + \langle p \rangle$ and so $Z_{i+1} \vDash_I I + \langle p \rangle$. Moreover, $Z_{i+1} \vDash_I \mathfrak{M}_i$. So $Z_{i+1} \vDash_I \langle \mathfrak{M}_i \rangle + I + \langle p \rangle$ and hence $Z_{i+1} \vDash_I \mathfrak{M}_{i+1}$. □

In the previous proof we did not exploit fully the fact that in the inductive hypothesis $\|H'\| \leqslant r - 4$. It is easy to adapt the previous proof to use this more carefully and indeed prove Theorem 4.3.

Restated Theorem 4.3 *Let V be a set of variables, $\mathbb{F}$ a field, $r \geqslant 1$ an integer, P an unsatisfiable set of polynomials and I a proper ideal in $\mathbb{F}[V]$. Let $P = P_1 \cup P_2$ and suppose that the following conditions hold:*

1. *there exists a non-empty (r,I)-BG family $\mathcal{F}$ for P_1;*
2. *for every $H \in \mathcal{F}$ s.t. $\|H\| < r$ and every $p \in P_2$ either $H \vDash_I p$ or there exists $H' = H \otimes H_p \in \mathcal{F}$, with $\|H_p\| = 1$, and there exists $\alpha \in H_p$, $p\!\restriction_\alpha \in I$.*

Then for every I-semantic polynomial calculus refutation π of P

$$\mathrm{MSp}(\pi) \geqslant \lfloor r/4 \rfloor \,. \tag{4.19}$$

Proof. The proof of Theorem 4.2 can be adapted here with minor (but non-trivial) modifications. Hence we use the same notations and, for sake of contradiction, we assume that $\mathrm{MSp}(\pi) < \lfloor r/4 \rfloor$.

The only difference with the proof of Theorem 4.2 is in the inductive step when $\mathfrak{M}_{i+1} \subseteq \langle \mathfrak{M}_i \rangle + I + \langle p \rangle$ with $p \in P_2$. By Lemma 4.1, used with parameters $H = H_i$, $Z = Z_i$ and $S = \mathfrak{M}_i$, we have a non-empty $H' \in \mathcal{F}$ and a non-empty 2-merge Z' of H' such that $Z' \vDash_I \mathfrak{M}_i$ and

$$\|H'\| \leqslant 4\mathrm{MSp}(\mathfrak{M}_i) \leqslant 4(r/4 - 1) \leqslant r - 4\,. \tag{4.27}$$

Then, by hypothesis (2) of the theorem, either $H' \vDash_I p$ or there is a flippable product H_p such that $H' \otimes H_p \in \mathcal{F}$ and $O_p = \{\alpha \in H_p \ : \ p\!\restriction_\alpha \in I\}$ is non-empty. In the first case just set $H_{i+1} = H'$ and $Z_{i+1} = Z'$. In the second case $\|H' \otimes H_p\| = \|H'\| + 1 < r$ so, again by hypothesis (2) of the theorem, either $H' \otimes H_p \vDash_I p$ or there exists a flippable product H'_p such that $H' \otimes H_p \otimes H'_p \in \mathcal{F}$ and $O'_p = \{\alpha \in H'_p \ : \ p\!\restriction_\alpha \in I\}$ is non-empty. Let $Z_p = (O_p \otimes H'_p) \cup (H_p \otimes O'_p)$. Then let $H_{i+1} = H' \otimes H_p \otimes H'_p$ and $Z_{i+1} = Z' \otimes Z_p$. These by construction satisfy the inductive hypothesis. $\square$

We can now turn to proving Lemma 4.1, but first we need a generalization of the concept of matchings in bipartite graphs. For the moment we focus on V-matchings; a further generalization will be given in Sect. 6.1.

Definition 4.7 (V-Matchings). Let G_V be a bipartite graph with three vertices shaped like a "V". More precisely it has vertices $\{v_0, v_1, v_2\}$ and two edges: $\{v_0, v_1\}$ and $\{v_1, v_2\}$. Its bipartition is $L(G_V) = \{v_1\}$ and $U(G_V) = \{v_0, v_2\}$.

Given a bipartite graph G with bipartition (L, U), a V-*matching* in G is a subgraph G' of G such that each connected component of G' is isomorphic to G_V by an isomorphism that maps $L(G_V)$ into L and $U(G_V)$ into U.[15]

The following is an easy corollary of Hall's theorem for matchings stated for reference in the Notation on p. xv.

Lemma 4.2 ([ABRW02]). *Let G be a bipartite graph with bipartition (L, U). The following are equivalent:*

1. *for each subset A of L, $|N(A)| \geqslant 2|A|$,*
2. *there exists a V-matching in G covering L.*

Proof. Clearly (2) implies (1). For the other implication, let G' be the auxiliary graph with bipartition (L', U) where each vertex $v \in L$ is "split" into two vertices $v_0, v_1 \in L'$

---

[15] This definition is essentially the same definition of V-matching in Sect. 6.1. The difference is that a V-matching in Sect. 6.1 could include singleton vertices from U. Hence, from the point of view of the vertices covered in L, the two notions are perfectly equivalent.

and (v_b, w) is an edge in G' if and only if (v, w) is an edge in G. Now, G' is such that for each subset A of L', $|N_{G'}(A)| \geqslant |A|$. Hence, by Hall's theorem, there exists a matching M covering L'. Then, merging the duplicated vertexes in L', from M we get a V-matching that covers L. □

We now use this simple combinatorial fact to prove Lemma 4.1. A visual hint for the notations used in this proof can be found in Fig. 4.2.

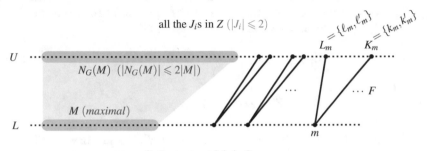

distinct monomials in S

Figure 4.2 A locality lemma

Proof (of Lemma 4.1). Let $H = H_1 \otimes \cdots \otimes H_t$ and $Z = Z_{J_1} \otimes \cdots \otimes Z_{J_r}$. Let G be the bipartite graph with bipartition (L, U) where L, the *lower* part of G, is indexed by the set of all *distinct* monomials in S, and U, the *upper* part of G, is indexed by the set $\{J_1, \ldots, J_r\}$. There is an edge (m, J_i) in G if and only if a variable of m appears in $\text{dom}(Z_{J_i})$. For a set $M \subseteq L$ let $N(M)$ be the set of neighbors of M in G and let H_M and Z_M be the following two flippable products:

$$Z_M = \bigotimes_{J_i \in N(M)} Z_{J_i}, \qquad H_M = \bigotimes_{J_i \in N(M)} \bigotimes_{j \in J_i} H_j. \qquad (4.28)$$

Let M be a set of maximal size in L such that $|N(M)| \leqslant 2|M|$. Let $M^c = L \setminus M$. By maximality of M, for each $A \subseteq M^c$, $|N(A) \setminus N(M)| \geqslant 2|A|$. Hence, by Lemma 4.2, there is a V-matching F covering M^c and F is in the subgraph of G induced by $M^c \cup (U \setminus N(M))$.

Let's start constructing H'. For each monomial $m \in M^c$, consider the upper part of the connected component F_m of F covering m and let $U(F_m) = \{L_m, K_m\}$ be this upper part, where $L_m, K_m \in \{J_1, \ldots, J_r\}$. By definition of G, there is a variable x in both m and $\text{dom}(Z_{L_m})$. Let $L_m = \{\ell_m, \ell'_m\}$ where ℓ_m is such that x is in $\text{dom}(H_{\ell_m})$. Analogously define $K_m = \{k_m, k'_m\}$. Define the product-family H' as

$$H' = H_M \otimes \bigotimes_{m \in M^c} H_{\ell_m} \otimes H_{k_m}. \qquad (4.29)$$

Clearly $H' \sqsubseteq H$ and hence it is a flippable product. The rank of H' is

$$\|H'\| = \|H_M\| + 2\,|M^c| \overset{(\star)}{\leqslant} 4\,|M| + 2\,|M^c| \leqslant 4\,|L| = 4\mathrm{MSp}(S)\,, \qquad (4.30)$$

where $(\star)$ follows from the fact that $|N(M)| \leqslant 2\,|M|$, and for all i, $|J_i| \leqslant 2$.

The construction of Z' goes as follows. Given $m \in M^c$ let

$$O_{m,i} = \{\alpha \in H_i \,:\, m{\restriction}_\alpha = 0\}\,. \qquad (4.31)$$

Observe that if a variable x of m is in $\mathrm{dom}(H_i)$ then $O_{m,i}$ is non-empty since H_i is flippable and hence there is always a Boolean assignment in H_i setting x to 0 and hence satisfying m, that is setting m to 0. As in Example 4.1, let

$$Z_{\{\ell_m,k_m\}} = (O_{m,\ell_m} \otimes H_{k_m}) \cup (H_{\ell_m} \otimes O_{m,k_m})\,. \qquad (4.32)$$

Then let Z' be

$$Z' = Z_M \otimes \bigotimes_{m \in M^c} Z_{\{\ell_m,k_m\}}\,. \qquad (4.33)$$

It is straightforward to see that Z' is a 2-merge on H'. We just have to show that $Z' \vDash_I S$. Since $Z \vDash_I S$, it is enough to show that for every Boolean assignment $\alpha \in Z'$ there is a Boolean assignment $\beta \in Z$ such that for every monomial m in S, $m{\restriction}_\alpha = m{\restriction}_\beta$. By the definition of Z' it is enough to show that for each $\alpha \in Z_{\{\ell_m,k_m\}}$ there exists $\beta \in Z_{L_m} \otimes Z_{K_m}$ such that for each monomial m in S, $m{\restriction}_\alpha = m{\restriction}_\beta$. Now, the fact that Z is a 2-merge means that $Z_{L_m} \supseteq H_{\ell_m} \otimes A_{\ell_m}$ for some non-empty $A_{\ell_m} \subseteq H_{\ell_m}$ and similarly $Z_{K_m} \supseteq H_{k_m} \otimes A_{k_m}$ for some non-empty $A_{k_m} \subseteq H_{k_m}$. Hence it is enough to take

$$\beta = \alpha \cup \bigcup_{m \in M^c} \alpha_{\ell'_m} \cup \alpha_{k'_m}\,, \qquad (4.34)$$

where $\alpha_{\ell'_m} \in A_{\ell'_m}$ and $\alpha_{k'_m} \in A_{k'_m}$. Then β is well defined since all the H_i are domain-disjoint, and it belongs to Z by construction. Now, given any monomial m in S, if $m \in M^c$ then $m{\restriction}_\alpha = 0$ and hence clearly $m{\restriction}_\beta = 0$. If $m \in M$ then by construction m does not contain variables from any of the sets $\mathrm{dom}(Z_{L_{m'}})$ and $\mathrm{dom}(Z_{K_{m'}})$ with $m' \in M^c$, hence, the Boolean assignment $\bigcup_{m \in M^c} \alpha_{\ell'_m} \cup \alpha_{k'_m}$ does not set variables in m. Then in both these cases $m{\restriction}_\alpha = m{\restriction}_\beta$. $\square$

4.6 Open Problems

Question 4.1. Let P be an unsatisfiable set of polynomials of degree at most d and let D be the minimum degree needed to refute P in polynomial calculus. Is it true that for every (semantic) polynomial calculus refutation π of P,

$$\mathrm{MSp}(\pi) = \Omega(D-d)\,? \qquad (4.35)$$

This question was asked for polynomial calculus (and more general proof systems) already in [ABRW02, Open question no. 4].

Regarding total space in polynomial calculus, all the questions in [ABRW02] are still open, in particular the following.

Question 4.2. Is there any family of CNF formulas F_n in n variables and with $n^{O(1)}$ clauses such that for every PCR refutation π of F_n

$$\text{TSp}(\pi) = \omega(n) \text{ ?} \tag{4.36}$$

The reason to ask for F_n to have $n^{O(1)}$ clauses is that otherwise CT_n answers eq. (4.36). But this is somehow an unsatisfactory answer since CT_n has exponentially many clauses in n.

Question 4.3. Do (r,I)-BG families, see Definition 4.5, characterize any interesting complexity measure in polynomial calculus? Can they used be to prove degree (and hence size) lower bounds in polynomial calculus?

The reason behind this question is that (r,I)-BG families have some similarities with the w-AD and w-BK families, see Sect. 2.3.1, which characterize respectively width and asymmetric width in resolution. Hence it seems plausible for the (r,I)-BG families to characterize something interesting in polynomial calculus. Moreover, if the (r,I)-BG families could provide degree lower bounds this would give a, possibly easier, way to prove such lower bounds alternative to the techniques in [BI10, AR01, MN15].

History

The main technical difficulty of this chapter is Lemma 4.1, the *Locality Lemma*. This is a generalization of [ABRW02, Lemma 4.14] and an analoguous result in [FLN$^+$15]. The way we present it is based on [BG15] although a preliminary equivalent version was already proved in [BG13].

The main definition of this chapter, the (r,I)-BG families (Definition 4.5), was introduced in [BG13] and simplified to this version in [BG15]. Theorem 4.3 is a generalization of [BG15, Theorem 3.6].

Part II
Applications

Chapter 5
Pigeonhole Principles

The *pigeonhole principle* asserts that there is no multi-valued total injective mapping from a set with m elements into a set with n elements, if $m > n$. The elements of the set of size m are traditionally called *pigeons* and the elements of the set of size n are called *holes* and so the pigeonhole principle can be stated more pictorially saying that

if $m > n$ pigeons fly to n holes then (at least) two of them must go to the same hole.

Interestingly, the proof complexity of the pigeonhole principle essentially depends on the number of pigeons m (as a function of the number of holes n) and on some details of its encodings as an unsatisfiable CNF formula or as an unsatisfiable set of polynomials. The encodings of the pigeonhole principle as an unsatisfiable CNF formula that we consider are the following:

- the "standard" pigeonhole principles, PHP_n^m, fPHP_n^m, oPHP_n^m, see Sect. 5.1;
- the "bit" pigeonhole principle, bitPHP_n^m, see Sect. 5.2;
- the "XOR" pigeonhole principle, xorPHP_n^m, see Sect. 5.3.

The *graph pigeonhole principle*, G-PHP, is instead considered much later in Sect. 7.3 since the monomial space lower bound for it relies on results from Chap. 6.

For each of the encodings above we review some of their properties with a focus on their space complexity in resolution and polynomial calculus. Indeed most of the space lower bounds from this section were already known in the literature [ABRW02, FLM$^+$13] but here we give short, self-contained proofs of those results as easy applications of the general theorems we saw in Part I.

5.1 The (Standard) Pigeonhole Principles

Let $m, n \in \mathbb{N}$ be two integers such that $m > n$ and consider the set of mn variables $X = \{x_{ij} : i \in [m], j \in [n]\}$. The intended meaning of x_{ij} is the truth value of "*the i-th pigeon goes into the j-th hole*". The standard encoding of the pigeonhole principle,

© Springer International Publishing AG, part of Springer Nature 2017
I. Bonacina, *Space in Weak Propositional Proof Systems*,
https://doi.org/10.1007/978-3-319-73453-8_5

PHP_n^m, asserts that there is an injective multi-valued mapping from $[m]$ to $[n]$. It is the conjunction of the following clauses:

1. $\neg x_{ij} \vee \neg x_{i'j}$ for all $i \neq i' \in [m]$ and for all $j \in [n]$ (**Injectivity Axioms**);
2. $x_{i1} \vee x_{i2} \vee \cdots \vee x_{in}$ for all $i \in [m]$ (**Totality Axioms**).

Clearly PHP_n^m is unsatisfiable whenever $m > n$ and its proof complexity has been investigated in depth since Haken used it to prove the first (sub-)exponential lower bounds on size for resolution [Hak85]: every resolution refutation π of PHP_n^{n+1} is such that

$$S(\pi) \geqslant 2^{\Omega(n)} . \tag{5.1}$$

Notice that this is not truly exponential in the number of variables of PHP_n^{n+1} since this formulas has $\Theta(n^2)$ variables. Regarding the relation between m and n, intuitively the larger m is with respect to n the 'more contradictory' PHP_n^m is, and interestingly its proof complexity indeed depends on the number of pigeons m with some *qualitative* changes occurring when

$$m = n+1, 2n, n^2, \infty . \tag{5.2}$$

For example in [Raz04, Raz03] it is proved that for every resolution refutation π of $PHP_n^{n^2}$

$$S(\pi) \geqslant 2^{\Omega(n/\log n)} . \tag{5.3}$$

In general for every $m > n$ it holds that PHP_n^m needs resolution refutations of size at least $2^{\Omega(\sqrt[3]{n})}$ [Raz03].

Regarding the upper bounds, PHP_n^{n+1} has resolution refutations of size $O(n^3 2^n)$ [BP97, Lemma 1]. More generally, PHP_n^m has polynomial-size refutations in proof systems such as cutting planes, see [CCT87, GPT15], and Frege systems, see [Bus87]. On the other hand, constant-depth Frege proofs of the pigeonhole principle require exponential size, see [Ajt94, BIK$^+$92, KPW95a, PBI93]. We refer to [Raz01] for a survey on the proof complexity of the pigeonhole principle.

We consider PHP_n^m in polynomial calculus and for the convenience of the reader we recall its encoding as a set of polynomials $P_{PHP_n^m}$ in the ring $\mathbb{F}[X \cup \overline{X}]$:

$$P_{PHP_n^m} = \left\{ x_{ij} x_{i'j} : i \neq i' \in [m] \text{ and } j \in [n] \right\}$$

$$\cup \left\{ \prod_{j \in [n]} \bar{x}_{ij} : i \in [m] \right\}$$

$$\cup \left\{ x_{ij}^2 - x_{ij}, x_{ij} + \bar{x}_{ij} - 1 : i \in [m] \text{ and } j \in [n] \right\} . \tag{5.4}$$

Notice that $P_{PHP_n^m}$ contains polynomials of degree n. This make degree lower bounds trivial, hence sometimes in polynomial calculus an alternative encoding of PHP_n^m is used. This is an encoding of PHP_n^m as a set of small-degree polynomials where the polynomials $\left\{ \prod_{j \in [n]} \bar{x}_{ij} \right\}_{i \in [m]}$ in $P_{PHP_n^m}$ are substituted by $\left\{ \sum_j x_{ij} - 1 \right\}_{i \in [n]}$. To avoid confusion we call this version of the pigeonhole principle where large degree initial polynomials are substituted by linear polynomials linPHP_n^m. Considering linPHP_n^m instead of PHP_n^m makes sense when proving degree lower bounds but it trivially implies monomial space lower bounds, as already some of the polynomials in linPHP_n^m requires a large, i.e., linear, number of monomials. For every $m > n$, linPHP_n^m require polynomial calculus refutations of degree $\Omega(n)$, see [Raz98], and hence polynomial calculus refutations of size $2^{\Omega(n)}$, due to eq. (4.8).

Regarding the space complexity of the pigeonhole principle, we have that it does not depend on the number of pigeons. Every resolution refutation of PHP_n^m requires clause space at least n [ET01, ABRW02]. Moreover, since PHP_n^m is an n-semiwide formula, see Definition 3.1, by Theorem 3.7 then every *semantic* resolution refutation of PHP_n^m requires total space at least $\lfloor n/2 \rfloor^2$. This result was proved in [ABRW02, Corollary 5.7] and indeed the proof of Theorem 3.7 can be seen as a generalization of this result. It also holds that every semantic polynomial calculus refutation of PHP_n^m requires total space at least $\Omega(n^2)$ [ABRW02, Corollary 5.7].

The monomial space lower bound we prove holds for the so called *onto* version of the pigeonhole principle, oPHP_n^m, that is the conjunction of PHP_n^m with the following clauses:

$$x_{1j} \vee x_{2j} \vee \cdots \vee x_{mj}, \tag{5.5}$$

for all $j \in [n]$ (**Onto Axioms**). We recall that the encoding of oPHP_n^m as a set of polynomials $tr(\mathsf{oPHP}_n^m)$ in the ring $\mathbb{F}[X \cup \overline{X}]$ is the following

$$P_{\mathsf{oPHP}_n^m} = P_{\mathsf{PHP}_n^m} \cup \left\{ \prod_{i \in [m]} \bar{x}_{ij} \right\}_{j \in [n]}. \tag{5.6}$$

Clearly oPHP_n^m is weaker than PHP_n^m, in the sense that it is even "more false" than PHP_n^m. Indeed any refutation of $P_{\mathsf{PHP}_n^m}$ is also a refutation of $P_{\mathsf{PHP}_n^m}$, so we can just focus on proving lower bounds on the complexity of oPHP_n^m.

Theorem 5.1 ([ABRW02]). *Let $\mathbb{F}$ be a field, $m > n$ two integers and let $X = \{x_{ij} : i \in [m] \text{ and } j \in [n]\}$. Let P_1 be the set of polynomials in $\mathbb{F}[X \cup \overline{X}]$ encoding the injectivity axioms of oPHP_n^m and the Boolean axioms, that is*

$$P_1 = \left\{ x_{ij} x_{i'j}, \ x_{ij}^2 - x_{ij}, \ x_{ij} + \bar{x}_{ij} - 1 \ : \ i \neq i' \text{ and } x_{ij}, x_{i'j} \in X \right\}. \tag{5.7}$$

Let $I = \langle P_1 \rangle$ be the ideal in $\mathbb{F}[X \cup \overline{X}]$ generated by P_1. Then for every I-semantic polynomial calculus refutation π of oPHP_n^m, it holds that

$$\mathrm{MSp}(\pi) \geqslant \lfloor n/4 \rfloor. \tag{5.8}$$

Proof. We apply Theorem 4.3. Let P be the polynomial encoding of oPHP_n^m, then $P = P_1 \cup P_2$ where P_1 is described above and

$$P_2 = \left\{ \prod_{j \in [n]} \bar{x}_{ij} \ : \ i \in [m] \right\} \cup \left\{ \prod_{i \in [m]} \bar{x}_{ij} \ : \ j \in [n] \right\}. \tag{5.9}$$

We want to construct a non-empty (n,I)-BG family $\mathcal{F}$ for P_1 satisfying the hypotheses of Theorem 4.3.

Given $i \in [m]$ and $j \in [n]$, let $(i \mapsto j)$ be the Boolean assignment with domain $\{x_{i'j}, \bar{x}_{i'j} : i' \in [m]\}$ defined as follows

$$(i \mapsto j)(x_{i'j}) = \begin{cases} 1 & \text{if } i' = i, \\ 0 & \text{if } i' \neq i, \end{cases} \tag{5.10}$$

$$(i \mapsto j)(\bar{x}_{i'j}) = \begin{cases} 0 & \text{if } i' = i, \\ 1 & \text{if } i' \neq i. \end{cases} \tag{5.11}$$

Let $H_j = \{(i \mapsto j) : i \in [m]\}$. Clearly H_j is flippable. Let $\mathcal{F}$ be the following family of flippable products: $H \in \mathcal{F}$ if and only if there exists a set of holes $A \subseteq [n]$ such that

$$H = \bigotimes_{i \in A} H_i. \tag{5.12}$$

It is immediate to check that $\mathcal{F}$ is a (r,I)-BG family for P_1. The only non-trivial property is the consistency property for the injectivity axioms but it is easy to see that each $H \in \mathcal{F}$ either sets polynomials $x_{ij}x_{i'j}$ to 0 or maps them to themselves. Let's see that P_2 satisfies the hypotheses of Theorem 4.3. Let $H \in \mathcal{F}$ with $\|H\| < n$ and let $p \in P_2$. Say that $p = \prod_{i \in [m]} \bar{x}_{ij}$. Since $\|H\| < n$ there exists a variable x_{ij} not set by assignments in H. Consider then $H \otimes H_j$. It belongs to $\mathcal{F}$ by construction and clearly there is an assignment $\alpha \in H_j$ such that $p{\restriction}_\alpha = 0 \in I$. The case $p = \prod_{j \in [n]} \bar{x}_{ij}$ is analogous. $\square$

Another standard even weaker pigeonhole principle is the *functional* one, fPHP$_n^m$, which is the conjunction of the PHP$_n^m$ formula with the following clauses

$$\neg x_{ij} \vee \neg x_{ij'}, \tag{5.13}$$

where $i \in [m]$ and $j, j' \in [n]$ are distinct (**Functionality Axioms**). As observed in [FLM$^+$13] the approach shown in Theorem 5.1 cannot give super-constant monomial space lower bounds for fPHP$_n^m$.

Regarding the total space in polynomial calculus we have the following lower bound.

Proposition 5.1 ([ABRW02, Theorem 5.1]). *Let $\mathbb{F}$ be a field, $X = \{x_1, \ldots, x_n\}$ a set of variables and I the ideal generated by the Boolean axioms in $\mathbb{F}[X \cup \overline{X}]$. Every I-semantic polynomial calculus refutation π of PHP$_n^m$ is such that*

$$\mathrm{TSp}(\pi) \geqslant \frac{3}{64}n^2. \square \tag{5.14}$$

5.2 The Bit-Pigeonhole Principle

Let $n = 2^k$ for $k \in \mathbb{N}$. The *bit* pigeonhole principle on n holes, bitPHP$_n$, is an unsatisfiable CNF formula over the variables $X = \{x_{ij} : i \in [n+1], j \in [k]\}$. It asserts that for all distinct $i, i' \in [n+1]$, the length-k binary strings $x_{i1} \ldots x_{ik}$ and $x_{i'1} \ldots x_{i'k}$ are distinct. We think of each element of $[n+1]$ as a pigeon and of the string $x_{i1} \ldots x_{ik}$ as the address, in binary, of the hole in $[n]$ that pigeon i is flying to. Understood in this way, bitPHP$_n$ asserts that there is an injective mapping of $n+1$ pigeons into n holes. Formally the principle consists of the clauses $B_h^{i,i'}$

$$B_h^{i,i'} = \bigvee_{j=1}^{k} (x_{ij} \not\equiv h_j) \vee (x_{i'j} \not\equiv h_j), \tag{5.15}$$

for each $i, i' \in [n+1]$ with $i < i'$ and each $h \in [n]$ such that its binary expansion is $h_1 \ldots h_k \in \{0,1\}^k$. The expression $x_{ij} \not\equiv h_j$ is a shortcut for $\neg x_{ij}$ if $h_j = 1$ and for x_{ij} if $h_j = 0$.

Then bitPHP$_n$ is a formula over $O(n \log n)$ variables consisting of $O(n^3)$ clauses each of width $O(\log n)$. Two motivations to study, and sometimes prefer, bitPHP$_n$ rather than PHP$_n^{n+1}$ are that its encoding is more efficient from the point of view of the number of variables used and that the width of its clauses is $O(\log n)$ instead of n.

Some of the properties of the usual PHP$_n^{n+1}$ carry over for bitPHP$_n$; for example it is easy to show a width lower bound for bitPHP$_n$ by constructing a $\Omega(n)$-AD family of assignments (see Definition 2.3). Usually it is the case that bitPHP$_n$ is harder than PHP$_n^{n+1}$. For instance PHP$_n^{n+1}$ has polynomial-size cutting planes refutations, while bitPHP$_n$ is hard for cutting planes: recently [HP17] proved that every cutting planes refutation of bitPHP$_n$ requires size at least $2^{\Omega(n^{1/8})}$.

Theorem 5.2 ([FLN$^+$15]). *Let $\mathbb{F}$ be a field, $n = 2^k$ an integer and let X be the set of variables $\{x_{ij} : i \in [n+1]$ and $j \in [k]\}$. Let I be the ideal in $\mathbb{F}[X \cup \overline{X}]$ generated by the Boolean axioms, $\{x_{ij}^2 - x_{ij}, \ x_{ij} + \bar{x}_{ij} - 1 : i \neq i'$ and $x_{ij}, x_{i'j} \in X\}$. Then for every I-semantic polynomial calculus refutation π of bitPHP$_n$ it holds that*

$$\mathrm{MSp}(\pi) \geqslant \lfloor n/8 \rfloor. \tag{5.16}$$

Proof. We use Theorem 4.2 and hence we just need to construct an $(n/2, I)$-BG family $\mathcal{F}$ for bitPHP$_n$.

Given a hole h with binary representation $(h_1, \ldots, h_k)_2$, let $\bar{h}$ be the hole with complementary binary representation $(1 - h_1, \ldots, 1 - h_k)_2$. Notice that if $h \in [n/2]$ then h and $\bar{h}$ are distinct. Given a set of holes A, let $\overline{A} = \{\bar{h} : h \in A\}$.

The notation $[i \mapsto h, i' \mapsto \bar{h}]$ where $i, i' \in [n+1]$ and $h \in [n]$ is short for the Boolean assignment with domain $\{x_{ij}, \bar{x}_{ij}, x_{i'j}, \bar{x}_{i'j} : j \in [k]\}$ where for each $j \in [k]$

$$\left[i \mapsto h, i' \mapsto \bar{h}\right](x_{ij}) = h_j \,, \tag{5.17}$$

$$\left[i \mapsto h, i' \mapsto \bar{h}\right](\bar{x}_{ij}) = 1 - h_j \,, \tag{5.18}$$

$$\left[i \mapsto h, i' \mapsto \bar{h}\right](x_{i'j}) = 1 - h_j \,, \tag{5.19}$$

$$\left[i \mapsto h, i' \mapsto \bar{h}\right](\bar{x}_{i'j}) = h_j \,. \tag{5.20}$$

Given $h \in [n/2]$ and an injective mapping $\sigma : \{h, \bar{h}\} \to [n+1]$, let H_h^{σ} be the set of Boolean assignments of domain $\left\{x_{\sigma(h)j}, x_{\sigma(\bar{h})j}, \bar{x}_{\sigma(h)j}, \bar{x}_{\sigma(\bar{h})j} : j \in [k]\right\}$:

$$H_h^{\sigma} = \left\{\left[\sigma(h) \mapsto h, \sigma(\bar{h}) \mapsto \bar{h}\right], \quad \left[\sigma(h) \mapsto \bar{h}, \sigma(\bar{h}) \mapsto h\right]\right\} \,. \tag{5.21}$$

By construction all the Boolean assignments in H_h^{σ} have the same domain, and H_h^{σ} is flippable and I-consistent. Consider then the family $\mathcal{F}$ of flippable products: $H \in \mathcal{F}$ if and only if there exists a set of holes $A \subseteq [n/2]$ and there exists an injective mapping $\sigma : A \cup \bar{A} \to [n+1]$ such that

$$H = \bigotimes_{h \in A} H_h^{\sigma} \,. \tag{5.22}$$

We prove that $\mathcal{F}$ is an $(n/2, I)$-BG family for bitPHP$_n$. The consistency and restriction property are obvious so we focus on the extension properties.

Let $H = \bigotimes_{h \in A} H_h^{\sigma} \in \mathcal{F}$ such that $\|H\| < n/2$ and consider p the polynomial encoding of $B_{i,i'}^h$ in $\mathbb{F}[X \cup \overline{X}]$. If both $i, i' \in \sigma(A \cup \bar{A})$ then, by construction, $H \vDash_I p$ and we have nothing to do. Otherwise, without loss of generality, assume $i' \notin \sigma(A \cup \bar{A})$. Since we have that $\|H\| = |A| < n/2$, there is some hole $h' \in [n/2] \setminus A$ and an injective σ' such that $\sigma' = \sigma \cup \{h' \mapsto i'\} \cup \{\bar{h}' \mapsto i''\}$ with i'' outside $\sigma(A \cup \bar{A}) \cup \{i'\}$. If $i \notin \sigma(A \cup \bar{A})$ take $i'' = i$. Let $H' = H \otimes H_{h'}^{\sigma'} = \bigotimes_{h \in A'} H_h^{\sigma'} \in \mathcal{F}$, where $A' = A \cup \{h'\}$. The family H' is clearly I-consistent and flippable and moreover $H' \vDash_I p$, as each assignment in H' sets i and i' to go into two distinct holes. More precisely, if $i \in \sigma(A \cup \bar{A})$ then i goes somewhere inside $A \cup \bar{A}$ and i' goes either in h' or $\bar{h}'$. If $i \in \sigma(A \cup \bar{A})$ take any $i'' \notin \sigma(A \cup \bar{A})$ and repeat the previous construction. Since i goes in $\bar{h}'$ and i' goes in h' or vice-versa we have in this case too that the extension $H' \vDash_I p$. □

Theorem 5.3 ([BGT14]). *Let π be a resolution refutation of* bitPHP$_n$, *then*

$$\mathrm{TSp}(\pi) = \Omega\left(n^2\right) \,. \tag{5.23}$$

Proof. It is immediate to see that the $(n/2, I)$-BG family for bitPHP$_n$ in the proof of Theorem 5.2 is also a $(n/2, 0)$-BG family for bitPHP$_n$. Hence Proposition 4.5 immediately gives the resolution total space lower bound. □

Alternatively, this total space lower bound can also be proved by proving a resolution width lower bound for the refutations of bitPHP$_n$, for instance by constructing a $\Omega(n)$-AD family for bitPHP$_n$ and then using Theorem 3.6.

Since bitPHP$_n$ has only $O(n \log n)$ variables, then the previous result is a total space lower bound in resolution that is super-linear in the number of variables. This result was indeed the very first super-linear total space lower bound for a formula with just polynomially many clauses.

5.3 The XOR Pigeonhole Principle

Let $m, n \in \mathbb{N}$ be two integers such that $m > n$ and consider the set of variables $X = \{x_{i,j} : i \in [m], j \in [n] \cup \{0\}\}$. A pigeon $i \in [m]$ is considered assigned to a hole $j \in [n]$ when $x_{i,j-1} \not\equiv x_{i,j}$ is true. The *XOR-Pigeonhole Principle*, xorPHP$_n^m$, expresses the following weaker form of the pigeonhole principle: if each pigeon is assigned to an odd number of holes, then there exists a hole with at least two pigeons. The formula xorPHP$_n^m$ is an unsatisfiable 4-CNF formula encoding the negation of this combinatorial principle. It is the conjunction of the following CNF formulas:

1. for each $i \in [m]$, $x_{i,0} \not\equiv x_{i,n}$, that is

$$(x_{i,0} \vee x_{i,n}) \wedge (\neg x_{i,0} \vee \neg x_{i,n});   \tag{5.24}$$

2. for all distinct $i, i' \in [m]$ and all $j \in [n] \cup \{0\}$,

$$(x_{i,j-1} \equiv x_{i,j}) \vee (x_{i',j-1} \equiv x_{i',j}),   \tag{5.25}$$

that is

$$(x_{i,j-1} \vee \neg x_{i,j} \vee x_{i',j-1} \vee \neg x_{i',j}) \wedge (\neg x_{i,j-1} \vee x_{i,j} \vee \neg x_{i',j-1} \vee x_{i',j})$$
$$\wedge (x_{i,j-1} \vee \neg x_{i,j} \vee \neg x_{i',j-1} \vee x_{i',j}) \wedge (\neg x_{i,j-1} \vee x_{i,j} \vee x_{i',j-1} \vee \neg x_{i',j}).   \tag{5.26}$$

Theorem 5.4 ([FLN$^+$15]). *Let $\mathbb{F}$ be a field, $m > n$ two integers and let X be the set of variables $\{x_{ij} : i \in [m] \text{ and } j \in [n]\}$. Let I be the ideal in $\mathbb{F}[X \cup \overline{X}]$ generated by the Boolean axioms, $\{x_{ij}^2 - x_{ij}, x_{ij} + \bar{x}_{ij} - 1 : i \neq i' \text{ and } x_{ij}, x_{i'j} \in X\}$. Then for every I-semantic polynomial calculus refutation π of xorPHP$_n^m$ it holds that*

$$\mathrm{MSp}(\pi) \geqslant \lfloor (n-1)/4 \rfloor .   \tag{5.27}$$

Proof. We use Theorem 4.2. Hence we just have to construct an $((n-1), I)$-BG family for xorPHP$_n^m$. Given $i \in [m]$ and $j \in [n]$, let $(i \mapsto j)$ and $(i \mapsto j)^*$ be the following two Boolean assignments of domain $\{x_{ij'}, \bar{x}_{ij'} : j' \in [n] \cup \{0\}\}$: for each $j' \in [n] \cup \{0\}$

$$(i \mapsto j)(x_{ij'}) = \begin{cases} 1 & \text{if } j' < j, \\ 0 & \text{if } j' \geqslant j, \end{cases} \qquad (5.28)$$

$$(i \mapsto j)(\bar{x}_{ij'}) = \begin{cases} 0 & \text{if } j' < j, \\ 1 & \text{if } j' \geqslant j, \end{cases} \qquad (5.29)$$

$$(i \mapsto j)^*(x_{ij'}) = \begin{cases} 0 & \text{if } j' < j, \\ 1 & \text{if } j' \geqslant j, \end{cases} \qquad (5.30)$$

$$(i \mapsto j)^*(\bar{x}_{ij'}) = \begin{cases} 1 & \text{if } j' < j, \\ 0 & \text{if } j' \geqslant j. \end{cases} \qquad (5.31)$$

Then for each i, j let

$$H_{i \mapsto j} = \{(i \mapsto j), (i \mapsto j)^*\} . \qquad (5.32)$$

By construction $H_{i \mapsto j}$ is flippable and I-consistent. Define $\mathcal{F}$ such that $H \in \mathcal{F}$ if and only if there exists a set $A \subseteq [m]$ of size at most $n-1$ and there exists an injective mapping $\mu : A \longrightarrow [n]$ such that

$$H = \bigotimes_{i \in A} H_{i \mapsto \mu(i)} . \qquad (5.33)$$

We prove that $\mathcal{F}$ is an $((n-1), I)$-BG family for xorPHP$_n^m$. The fact that $\mathcal{F}$ is non-empty, each $H \in \mathcal{F}$ is I-consistent and the restriction property are very easy to check so we focus on proving the extension property of $\mathcal{F}$.

Let $H = \bigotimes_{i \in A} H_{i \mapsto \mu(i)} \in \mathcal{F}$ with $\|H\| < n-1$ and p the polynomial encoding of an initial clause C from xorPHP$_n^m$. Let us suppose first that C is a clause from some $(x_{i,j-1} \equiv x_{i,j}) \vee (x_{i',j-1} \equiv x_{i',j})$. If both i and i' are in A, then, by construction, $H \vDash_I p$ and we have no extension to perform. If $i \notin A$, then, as μ is an injective assignment of at most $n-2$ pigeons, we can find a hole h *different from* j which is not in $\mu(A)$. Then let $\mu' = \mu \cup \{i \mapsto h\}$ and $H' = \bigotimes_{\ell \in A \cup \{i\}} H_{\ell \mapsto \mu'(\ell)} = H \otimes H_{i \mapsto h}$. By construction $H_{i \mapsto h} \vDash_I p$, hence $H' \vDash_I p$. Similarly if $C = (x_{i,0} \not\equiv x_{i,n})$ we proceed as before extending μ to assign the pigeon i somewhere (if needed). $\square$

Theorem 5.5. *Let π be a resolution refutation of* xorPHP$_n^m$*, then*

$$\mathrm{TSp}(\pi) = \Omega\left(n^2\right) . \qquad (5.34)$$

Proof. It is immediate to see that the $(n/2, I)$-BG family for xorPHP$_n$ in the proof of Theorem 5.4 is also an $(n/2, 0)$-BG family for xorPHP$_n$. Hence Proposition 4.5 immediately gives the resolution total space lower bound. $\square$

Alternatively, this total space lower bound can also be proved by proving a resolution width lower bound for the refutations of bitPHP$_n$, for instance constructing an $\Omega(n)$-AD family for bitPHP$_n$ and then using Theorem 3.6.

Notice that unlike the analogous theorem for bitPHP$_n$, this total space lower bound for xorPHP$_n$ is just linear in the number of variables and not super-linear as it is for bitPHP$_n$.

5.4 Open Problems

Question 5.1. Is it true that for every semantic polynomial calculus refutation π of fPHP$_n^m$, it holds that

$$\mathrm{MSp}(\pi) = \Omega(n) \ ? \tag{5.35}$$

In [FLM$^+$13] it is observed that the technique of (r, I)-BG families will not prove super-linear monomial space lower bounds. More generally we have the following related open question.

Question 5.2. Is it true that if F is r-semiwide then for every polynomial calculus refutation π of F

$$\mathrm{MSp}(\pi) = \Omega(r) \ ? \tag{5.36}$$

Then as particular cases of this question we have the previous question about monomial space lower bounds for the functional pigeonhole principle and an analogous question for *Graph Tautologies*, see [ABRW02, Definition 3.12].

History

More direct, essentially equivalent, proofs of the monomial lower bounds for bitPHP$_n$ and xorPHP$_n$ can be found in [FLM$^+$13]. A more direct proof of the total space lower bound for bitPHP$_n$ can be found in [BGT16].

Chapter 6
Interlude: Cover Games

Let's put aside proof complexity for a moment and, in this chapter, we will focus on some properties of bipartite graphs. We start by generalizing the concept of *matchings* in bipartite graphs to what we call $\mathcal{C}$-*matchings*. Intuitively a $\mathcal{C}$-*matching* in a bipartite graph G with bipartition $(L(G), U(G))$ is a collection of vertex-disjoint subgraphs of G isomorphic to some graph from the collection of graphs $\mathcal{C}$. Our main interest is in two particular cases of $\mathcal{C}$-matchings: the V-*matchings* and the VW-*matchings*. We already saw a version of the V-matchings in Sect. 4.5.2, and the VW-matchings are just particular $\mathcal{C}$-matchings in which each connected component looks like a 'V', a 'W' or a singleton from $U(G)$, see Sect. 6.1 for the formal definitions.

First we prove a version of Hall's theorem that holds for VW-matchings, see Theorem 6.1. Then, in Sect. 6.1, we define some general two-player games, the *Cover Games*, that generalize a *matching game* played on bipartite graphs [BG03]. Both the Cover Games and the matching game are played between two players, Choose and Cover. Informally, given a bipartite graph G, a winning strategy for Cover in the matching game guarantees that there is a family of matchings $\mathcal{F}$ such that each matching in $\mathcal{F}$ can be enlarged to cover new vertexes in G (chosen by Choose) or shrunk while remaining in $\mathcal{F}$ and the family $\mathcal{F}$ has large matchings in it. The *Cover Game* is this exact game but instead of using matching we use $\mathcal{C}$-matchings. We then show some necessary conditions for the existence of winning strategies for Cover in the Cover Game over V-matchings and VW-matchings, see resp. Theorem 6.2 and Theorem 6.3. Finally we show that there exists a winning strategy for Cover if the Cover Game is played on random bipartite graphs (using V-matchings or VW-matchings), see Theorem 6.5.

6.1 $\mathcal{C}$-Matchings

A $\mathcal{C}$-matching is a generalization of the usual notion of matchings in bipartite graphs and the V-matchings we saw in Sect. 4.5.2.

© Springer International Publishing AG, part of Springer Nature 2017
I. Bonacina, *Space in Weak Propositional Proof Systems*,
https://doi.org/10.1007/978-3-319-73453-8_6

Definition 6.1 (C-Matchings). Let C be a collection of bipartite graphs G_i with bipartition $(L(G_i), U(G_i))$ and G be a bipartite graph with bipartition $(L(G), U(G))$. A C-*matching* in G is a subgraph G' of G such that each connected component of G' is isomorphic to some graph G_j in C by an isomorphism that maps $L(G_j)$ into $L(G)$ and $U(G_j)$ into $U(G)$.

In what follows we are interested in C-matchings for particular collections of graphs $\{G_\bullet, G_V\}$ and $\{G_\bullet, G_V, G_W\}$. For simplicity we call the $\{G_\bullet, G_V\}$-matchings simply V-*matchings* and the $\{G_\bullet, G_V, G_W\}$-matchings simply VW-*matchings*. The bipartite graphs $G_\bullet$, G_V and G_W are defined as follows:

(a) $G_\bullet$ consists of a single vertex v and no edges. The vertex v belong to $U(G_\bullet)$.
(b) G_V has three vertices $\{v_0, v_1, v_2\}$ and two edges: $\{v_0, v_1\}$ and $\{v_1, v_2\}$. Its bipartition is $L(G_V) = \{v_1\}$ and $U(G_V) = \{v_0, v_2\}$. It has the shape of a V.
(c) G_W has five vertices $\{v_0, v_1, v_2, v_3, v_4\}$ and four edges: $\{v_0, v_1\}$, $\{v_1, v_2\}$, $\{v_2, v_3\}$ and $\{v_3, v_4\}$. Its bipartition is $L(G_W) = \{v_1, v_3\}$ and $U(G_W) = \{v_0, v_2, v_4\}$. It has the shape of a W.

Our main interest in V-matchings and VW-matchings is that they are among the simplest trees, and in some graphs associated with CNF formulas we can use such trees to build (r, I)-BG families of assignments, see Sect. 7.1, and ultimately space lower bounds in polynomial calculus.

6.2 Some Hall's Theorems

For V-matchings and VW-matchings we can prove versions of the usual Hall's theorem for matchings (see for reference the Notations section on page xv). That is we are looking for results relating the qualitative property of the existence of V-matchings and VW-matchings in bipartite graphs to the expansion properties of the graph.

In Sect. 4.5.2 we already saw one such result for V-matchings: the following result, which is an immediate consequence of Hall's theorem.

Restated Lemma 4.2 ([ABRW02]) *Let G be a bipartite graph with bipartition (L, U). The following are equivalent:*

1. *for each subset A of L, $|N(A)| \geqslant 2|A|$,*
2. *there exists a V-matching in G covering L.*

Here we want to prove an analogue of Hall's theorem and Lemma 4.2 for VW-matchings, but, before doing that, let's see why we can't get an exact analogue of such results. For convenience we recall here the notion of bipartite expansion, which is also in the Notations section.

Definition 6.2 (Bipartite Expansion). Let $r \in \mathbb{N}$ and $c \in \mathbb{R}$. A bipartite graph G with bipartition (L, U) is an (r, c)-*bipartite expander* if and only if

$$\forall A \subseteq L(G), |A| \leqslant r \rightarrow |N_G(A)| \geqslant c|A| \ . \tag{6.1}$$

Proposition 6.1. *Let G be a bipartite graph and let $|L(G)| = n$. If there exists a VW-matching in G covering $L(G)$ then G is an $(n, \frac{3}{2})$-bipartite expander.*

Proof. If G has as subgraph a VW-matching covering $L(G)$ then clearly G is an $(n, \frac{3}{2})$-bipartite expander since each subset of a VW-matching expands in G by a ratio of at least $\frac{3}{2}$.

Unfortunately, the converse of Proposition 6.1 does not hold, as the following example shows.

Example 6.1. An easy counterexample to the converse of Proposition 6.1 is the following graph D_n, see Fig. 6.1 for D_4.

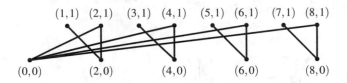

$(1,1)$ $(2,1)$ $(3,1)$ $(4,1)$ $(5,1)$ $(6,1)$ $(7,1)$ $(8,1)$

$(0,0)$ $(2,0)$ $(4,0)$ $(6,0)$ $(8,0)$

Figure 6.1 D_4

The graph D_n has vertex-set

$$([2n] \times \{1\}) \cup (\{0,2,4,\dots,2n\} \times \{0\}), \tag{6.2}$$

with $L(G) = \{0,2,4,\dots,2n\} \times \{0\}$ and $U(G) = [2n] \times \{1\}$. Its edges are all the

$$\{(0,0),(2i,1)\}, \{(2i,0),(2i,1)\}, \{(2i,0),(2i-1,1)\}, \tag{6.3}$$

for $1 \leqslant i \leqslant n$. We have that D_n is an $(n+1, \delta_n)$-bipartite expander where $\delta_n \rightarrow 2$ as $n \rightarrow \infty$. But on the other hand there is no VW-matching in D_n covering $L(D_n)$.

We want to show some kind of converse of Proposition 6.1 and the example we just saw tells us we have to proceed a bit carefully.

6.2.1 A Hall's Theorem for VW-Matchings

The following theorem behaves like a sort of converse of Proposition 6.1. The way we present it is somehow tailored to the applications we are interested in, that is bipartite graphs G such that each $v \in L(G)$ has degree at most 3.

Theorem 6.1 ([BBG$^+$17, Lemma 1.2]^1). *Let G be a bipartite graph. Suppose that the following properties hold:*

1. *for each $v \in L(G)$, $\deg(v) \leqslant 3$ and no pair of degree 3 vertices in $L(G)$ have the same set of neighbors,*
2. *$|N(L(G))| \geqslant c\,|L(G)|$, with $c > 1.8$ and*
3. *each proper subset of $L(G)$ can be covered by a VW-matching,*

then $L(G)$ can be covered by a VW-matching.

Proof. For sake of contradiction, let G be a bipartite graph witnessing the fact that the theorem is not true and for brevity let $L = L(G)$ and $U = U(G)$. Without loss of generality we can suppose that $U = N_G(L)$. By hypothesis, every proper subset of L can be covered by a VW-matching but the whole L cannot. This means that the graphs in Fig. 6.2 cannot be mapped into G by a mapping respecting the edge adjacencies, the degree of the vertexes and mapping the lower part of such graphs into L and the upper part into U.

(a)   (b)   (c)   (d)   (e)

○   node of arbitrary degree
●   node of degree fixed by the figure

Figure 6.2 List of forbidden subgraphs for G

We say that two vertices v, v' in U are *close* if there exists a vertex $w \in L$ such that $v, v' \in N(w)$. We now weight each vertex in U by its degree and we redistribute the weight in the following way: each vertex in U of degree 1 gets weight $\frac{1}{3}$ from its close vertices. Let v be a vertex in U and let $w(v)$ be its weight at the end of the previous process. Then, since we are just redistributing the weight,

$$\sum_{v \in U} \deg(v) = \sum_{v \in U} w(v) . \tag{6.4}$$

If $v \in U$ is such that $\deg(v) = 1$ then

$$w(v) = \begin{cases} 1 + 1/3 & \text{if } v \text{ has only one close vertex },\\ 1 + 2/3 & \text{otherwise }. \end{cases} \tag{6.5}$$

---

^{1} The original proof from [BBG$^+$17] used the assumption of $c > 1.96$. Susanna Figueiredo de Rezende (*pers. comm.*) later simplified the argument and gave a better bound ($c > 1.8$). With her kind permission we give here her simplified proof. For more details about this result see the History section at the end of this chapter.

In fact, since the graphs in Fig. 6.2.(a) and (b) are not subgraphs of G, we have that two vertices of degree 1 in U cannot be close.

If $v \in U$ is such that $\deg(v) = 2$ then

$$w(v) \geqslant 2 - 1/3 . \tag{6.6}$$

Indeed, since the graphs in Fig. 6.2. (c), (d) and (e) are not subgraphs of G, a vertex of degree 2 in U can be close to at most one vertex of degree 1 in U.

If $v \in U$ is such that $\deg(v) = d \geqslant 3$ then $w(v) \geqslant 2$, since it can be close to at most d vertices of degree 1 as the graph in Fig. 6.2.(b) is not a subgraph of G, and hence $w(v) \geqslant d - \frac{d}{3} = \frac{2}{3}d \geqslant 2$.

Let $L = L_2 \cup L_3$, where L_i are the vertices of degree i in L, and U' be the set of degree 1 vertices of U that have only one close vertex. This means that each $u \in U'$ is a neighbor of some vertex in L_2 and since no pair of vertices of degree 1 can be in the same neighborhood, then $|U'| \leqslant |L_2|$. Therefore we have the following chain of inequalities:

$$\frac{3}{c}|N(L)| \geqslant 3|L| = 3|L_3| + 2|L_2| + |L_2| = \sum_{v \in U} \deg(v) + |L_2| = \tag{6.7}$$

$$= \sum_{v \in U} w(v) + |L_2| \geqslant \frac{5}{3}|U| - \frac{1}{3}|U'| + |L_2| \geqslant \frac{5}{3}|U| = \tag{6.8}$$

$$= \frac{5}{3}|N(L)| . \tag{6.9}$$

It follows that $c \leqslant 9/5$, which contradicts the hypothesis on c. $\quad\square$

6.3 Cover Games

We now introduce some combinatorial games on bipartite graphs. These can be seen as a generalization of matching games on bipartite graphs [BG03].

Let G be a bipartite graph. Given a subgraph F of G and a subset A of vertices of G, we recall that F *covers* A if A is contained in the vertex-set of F.

Definition 6.3 (Cover Games). Let G be a bipartite graph, μ an integer and $\mathcal{C}$ a collection of bipartite graphs. The *Cover Game* $\text{CovGame}_{\mathcal{C}}(G, \mu)$ is a game on the bipartite graph G between two players called Choose (he) and Cover (she). At each step i of the game the players maintain a $\mathcal{C}$-matching F_i in G. They start with the empty $\mathcal{C}$-matching and at step $i+1$ Choose can either

1. remove a connected component from F_i, or
2. if the number of connected components of F_i is strictly less than μ, pick a vertex (either in $L(G)$ or $U(G)$) and challenge Cover to find a $\mathcal{C}$-matching F_{i+1} in G such that

   a. each connected component of F_i is also a connected component of F_{i+1};

      b. F_{i+1} covers the vertex picked by Choose.

Cover loses the game $\mathsf{CovGame}_\mathcal{C}(G,\mu)$ if at some point she cannot answer a challenge by Choose. If the game can go on indefinitely without Cover losing we say that she wins.

     We are interested in winning conditions for the player Cover for the cover games where V-matchings and VW-matchings are used, that is $\mathsf{CovGame}_\mathsf{V}(G,\mu)$ and $\mathsf{CovGame}_\mathsf{VW}(G,\mu)$. The reason for our interest is the fact that for suitable graphs G associated with sets of monomials a winning strategy for the player Cover for the cover game $\mathsf{CovGame}_\mathcal{C}(G,\mu)$ implies the existence of a (μ,I)-BG family for some ideal I (under some assumptions on $\mathcal{C}$) and hence ultimately some monomial space lower bound, see Sect. 7.1.

     The ultimate goal of this chapter is to prove that if G is a random bipartite graph then Cover has a winning strategy for $\mathsf{CovGame}_\mathsf{V}(G,\mu)$ and $\mathsf{CovGame}_\mathsf{VW}(G,\mu)$ for μ linear in the number of vertices of G. This will be proven in Sect. 6.4 as a consequence of the following two theorems showing the existence of a winning strategy for Cover when G is a bipartite expander (and it fulfills some more technical hypotheses).

Theorem 6.2. *Let G be a bipartite graph with bipartition $(L(G), U(G))$, r a positive integer and $c > 2$ a real number. Suppose that the following two properties hold:*

    1. G is an (r,c)-bipartite expander;
    2. the maximum degree of a vertex in $U(G)$ is at most μ.

Then Cover wins $\mathsf{CovGame}_\mathsf{V}(G,\mu)$ with $\mu = \frac{r(c-2)}{2d^2}$ where d is the maximum degree of a vertex in $L(G)$.

     For some applications the assumption that $c > 2$ in the previous theorem is not enough. Luckily an analogous theorem holds for slightly lower c, and this is enough for our applications.

Theorem 6.3. *Let G be a bipartite graph with bipartition $(L(G), U(G))$, r, D be positive integers, and $c > 1.9$ be a real number. For every integer $d \geqslant D$ let S_d be the set of vertices of $U(G)$ with degree bigger than d. Suppose that*

    1. each vertex in $L(G)$ has degree at most 3;
    2. G is an (r,c)-bipartite expander;
    3. for every $D_{\max} \geqslant d \geqslant D$,

$$r(2-c) \geqslant 72d(|S_d|+d),\tag{6.10}$$

where $D_{\max}$ is the maximum degree of a vertex in $U(G)$.

Then Cover wins $\mathsf{CovGame}_\mathsf{VW}(G,\mu)$ with $\mu = \frac{r(2-c)}{72D}$.

     The proofs of Theorem 6.2 and Theorem 6.3 are quite long and a bit technical. We prefer to focus first on their applications to random bipartite graphs (in the next section) and leave to Sect. 6.5 their full proofs.

6.4 Random Bipartite Graphs

A bipartite graph is an (n,d,Δ)-*random bipartite graph* if it is chosen according to the uniform distribution on the set of bipartite graphs G such that $U(G) = n$ and $L(G) = \Delta n$ and the degree of any vertex in $L(G)$ is d.

In this section we prove that with high probability Cover has winning strategies for Cover Games over random bipartite graphs. To prove this fact we need to check that with high probability random bipartite graphs fulfill all the hypotheses of Theorem 6.2 and Theorem 6.3. This is the content of the following theorem and the next two lemmas.

Theorem 6.4 ([BG03, Lemma 5.1]). *For any $d \geqslant 3$, $\Delta \geqslant 1$ and any real constant $c \in (1, d-1)$, there is a constant $\gamma = \gamma_{d,c,\Delta}$ such that, for large n, if G is an (n,d,Δ)-random bipartite graph then, with high probability, G is a $(\gamma n, c)$-bipartite expander.* □

The proof of the previous theorem is standard and can be found (in slightly different forms), for instance in [HLW06, CS88, BP96, BW01, BG03].

Lemma 6.1. *Let G be an (n,d,Δ)-random bipartite graph with Δ and d positive constants. Then, with high probability, there is no vertex in $U(G)$ of degree bigger than $\log n$.*

Proof. The expected number of vertices in $U(G)$ of degree at least $\log n$ is at most

$$n \binom{\Delta n}{\log n} \left(\frac{\binom{n-1}{d-1}}{\binom{n}{d}} \right)^{\log n} \leqslant n \left(\frac{e\Delta n}{\log n} \right)^{\log n} \left(\frac{d}{n} \right)^{\log n} = o(1). \tag{6.11}$$

Hence, with high probability, there are no such vertices. □

Lemma 6.2. *Let Δ be a constant and G be an $(n,3,\Delta)$-random bipartite graph. For every integer d, let $S_d = \{v \in U(G) : \deg_G(v) \geqslant d\}$. Then for every real constant $\delta > 0$, with high probability for sufficiently large n there exists a constant D such that for every $\log n \geqslant d \geqslant D$,*

$$d(|S_d| + d) \leqslant \delta n. \tag{6.12}$$

Proof. We claim that for every $\log n \geqslant d \geqslant 12e\Delta$, with high probability

$$|S_d| \leqslant \frac{en}{2^d}. \tag{6.13}$$

Before proving eq. (6.13), we show how to conclude the desired bound on S_d. Fix a positive constant δ and let $\log n \geqslant D \geqslant 12e\Delta$ be big enough that $\frac{eD}{2^D} \leqslant \delta/2$. Moreover, for sufficiently large n, we have also that $\log^2 n \leqslant \delta n/2$. For d such that $\log n \geqslant d \geqslant D$ we have the following chain of inequalities:

$$d(|S_d|+d) \overset{\text{eq. (6.13)}}{\leqslant} \frac{end}{2^d} + d^2 \leqslant \frac{eDn}{2^D} + \log^2 n \leqslant \frac{\delta n}{2} + \frac{\delta n}{2} = \delta n . \qquad (6.14)$$

It remains to show just equation (6.13). Consider $\log n \geqslant d \geqslant 12e\Delta$. The probability that there are at least $\frac{en}{2^d}$ vertexes in $U(G)$ of degree at least d is at most

$$\Pr\left[|S_d| \geqslant \frac{en}{2^d}\right] \leqslant \binom{n}{\frac{en}{2^d}}\left[\binom{\Delta n}{d}\left(\frac{3}{n}\right)^d\right]^{\frac{en}{2^d}} . \qquad (6.15)$$

$$\leqslant \left[2^d\left(\frac{e\Delta n}{d}\right)^d\left(\frac{3}{n}\right)^d\right]^{\frac{en}{2^d}} \qquad (6.16)$$

$$\leqslant \left(\frac{6e\Delta}{d}\right)^{\frac{edn}{2^d}} \qquad (6.17)$$

$$\leqslant \left(\frac{1}{2}\right)^{\frac{edn}{2^d}} \qquad (6.18)$$

$$= o(1) , \qquad (6.19)$$

where eq. (6.19) holds since $\log n \geqslant d \geqslant 12e\Delta$, and we used the standard estimation $\binom{n}{m} \leqslant \left(\frac{en}{m}\right)^m$. □

Given the previous lemmas and Theorem 6.2 and Theorem 6.3, then the proof of the existence of a winning strategy for Cover for the Cover Game on bipartite random graphs is quite straightforward.

Theorem 6.5. *Let $d \geqslant 3$, $\Delta \geqslant 1$ and G be an (n,d,Δ)-random bipartite graph. Then, for large n, with high probability there exists a constant γ such that Cover has a winning strategy for* CovGame$_{\text{VW}}(G,\gamma n)$. *Moreover, if $d \geqslant 4$ then Cover has a winning strategy for* CovGame$_{\text{V}}(G,\gamma n)$.

Proof. Consider separately the case of $d \geqslant 4$ and $d = 3$. In the first case pick any constant $c \in (2,3)$, e.g., $c = 2.5$. Then, by Theorem 6.4, with high probability for large n there exists a constant $\gamma = \gamma_{d,c,\Delta}$ such that G is a $(\gamma n, c)$-bipartite expander. Moreover, by Lemma 6.1, no vertex in $U(G)$ has degree bigger than $\log n$ and henceforth, for large n, no vertex in $U(G)$ has degree bigger than γn. Hence, for large n, with high probability, G satisfies the hypotheses of Theorem 6.2, hence Cover has a winning strategy for CovGame$_{\text{V}}(G,\gamma n)$.

In the case of $d = 3$ pick $c \in (1.9,2)$, e.g., $c = 1.95$. Then, by Theorem 6.4, with high probability for large n there exists a constant $\gamma' = \gamma'_{c,\Delta}$ such that G is a $(\gamma' n, c)$-bipartite expander. Moreover, with high probability, by Lemma 6.1, the maximum degree of a vertex in $U(G)$ is $\log n$ and, by Lemma 6.2, for large enough n there exists a constant D such that for every $\log n \geqslant d \geqslant D$,

$$72d(|S_d|+d) \leqslant (2-c)\gamma' n , \qquad (6.20)$$

where $S_d = \{v \in U(G) : \deg_G(v) \geqslant d\}$. Hence, for large n, with high probability G satisfies the hypotheses of Theorem 6.3, and then Cover has a winning strategy for CovGame$_{VW}(G, \gamma n)$ with $\gamma = \frac{(2-c)\gamma'}{72D}$. $\quad\square$

6.5 Winning Strategies for Cover Games

This last part of the chapter is more technical and is dedicated to proving Theorem 6.2 and Theorem 6.3. The structure of both proofs is very similar but there are some technical differences. We start with the proof of Theorem 6.2 since it is simpler.

6.5.1 A Winning Strategy for the Game on V-Matchings

To simplify the exposition in this subsection we consider a fixed bipartite graph G, an integer $r \geqslant 1$ and a real number $c > 2$ such that G is an (r,c)-bipartite expander.

For brevity let $L = L(G)$, $U = U(G)$ and d be the maximum degree of a vertex in L. Given $A \subseteq L$ and $B \subseteq U$, we let $G_{A,B}$ be the subgraph of G induced by $(L \cup U) \setminus (A \cup B)$.

Definition 6.4 (V-Matching Property). Given two sets $A \subseteq L$ and $B \subseteq U$, we say that the pair (A, B) has the V-*matching property* if for every $C \subseteq L \setminus A$ with $|C| \leqslant r$, there exists a V-matching F in $G_{A,B}$ covering C.

Lemma 6.3. *Let $A \subseteq L$ and $B \subseteq U$ be such that the pair (A, B) does not have the* V-*matching property. Then there exists a set $C \subseteq L \setminus A$ with $(c-2)|C| < |B|$, such that no* V-*matching in $G_{A,B}$ covers C.*

Proof. Take $C \subseteq L \setminus A$ of minimal size such that no V-matching in $G_{A,B}$ covers C. We have that $|C| \leqslant r$. By the minimality of C, Lemma 4.2 implies that

$$\left| N_{G_{A,B}}(C) \right| < 2|C| . \tag{6.21}$$

By hypothesis, G is an (r,c)-bipartite expander. Hence $c|C| \leqslant |N_G(C)|$. Therefore,

$$c|C| \leqslant |N_G(C)| \leqslant \left| N_{G_{A,B}}(C) \right| + |B| < 2|C| + |B| . \tag{6.22}$$

Hence $(c-2)|C| < |B|$. $\quad\square$

Lemma 6.4. *The pair $(\emptyset, \emptyset)$ has the* V-*matching property.*

Proof. By contradiction suppose that $(\emptyset, \emptyset)$ does not have the V-matching property. Then, by Lemma 6.3, there exists a set $C \subseteq L \setminus A$ that has no V-matching in $G_{A,B}$ covering C and C has *negative* size, which is clearly not possible. $\quad\square$

Lemma 6.5 (Component Removal). *Let $A \subseteq L$ and $B \subseteq U$ be such that the pair (A,B) has the V-matching property and*

$$(c-2)r \geqslant |B| \ . \tag{6.23}$$

Then for each V-matching F contained in the subgraph of G induced by $A \cup B$, we have that $(A \setminus L(F), B \setminus U(F))$ has the V-matching property.

A visual hint for the notations used in this proof can be found in Fig. 6.3.

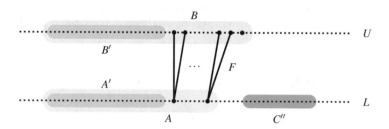

Figure 6.3 Component removal for V-matchings

Proof. Let $A' = A \setminus L(F)$ and $B' = B \setminus U(F)$ and suppose, for a contradiction, that (A',B') does not have the V-matching property. By Lemma 6.3, it is sufficient to prove that for each set $C \subseteq L \setminus A'$ with $(c-2)|C| < |B'|$, there is a V-matching in $G_{A',B'}$ covering C. Let $C' = C \cap L(F)$ and $C'' = C \setminus C'$. By construction, F is a V-matching such that $L(F) \subseteq A$, $U(F) \subseteq B$ and F covers C'. Moreover, we have that

$$|C''| \leqslant |C| < \frac{1}{(c-2)} |B'| < \frac{1}{(c-2)} |B| \overset{\text{eq. (6.23)}}{\leqslant} r \ . \tag{6.24}$$

Hence there exists a V-matching F'' of C'' in $G_{A,B}$, and since F and F'' are vertex-disjoint, then $F \cup F''$ is a V-matching in $G_{A',B'}$. By construction $F \cup F''$ covers C. □

Lemma 6.6 (Covering a Vertex in L). *Let $A \subseteq L$ and $B \subseteq U$ be such that the pair (A,B) has the V-matching property. If*

$$2(c-2)r \geqslant d^2(|B|+2) \ , \tag{6.25}$$

then for each vertex v in $L \setminus A$, there exists a V-matching F in $G_{A,B}$ covering v and such that $(A \cup L(F), B \cup U(F))$ has the V-matching property.

A visual hint for the notations used in this proof can be found in Fig. 6.4.

Proof. Fix $v \in L \setminus A$ and let S be the set of all V-matchings F in $G_{A,B}$ covering v and such that F has a single connected component, that is F is isomorphic to the graph G_V from item (b) on p. 72.

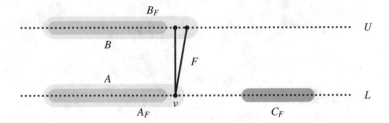

Figure 6.4 Covering a vertex in L via V-matchings

Since $r \geqslant 1$ and (A,B) has the V-matching property, then S is non-empty. For every $F \in S$, let (A_F, B_F) be the pair $(A \cup L(F), B \cup U(F))$, and suppose, for sake of contradiction, that for every $F \in S$, the pair (A_F, B_F) does not have the V-matching property. By Lemma 6.3, for every $F \in S$ there exists a set $C_F \subseteq L \setminus A_F$ with $|C_F| < \frac{1}{(c-2)}|B_F|$ and such that there is no V-matching of C_F in G_{A_F, B_F}.

Let $C = \bigcup_{F \in S} C_F$. Then we have the following chain of inequalities

$$|C| \leqslant \sum_{F \in S} |C_F| \tag{6.26}$$

$$< \frac{1}{(c-2)} \sum_{F \in S} |B_F| \tag{6.27}$$

$$\leqslant \frac{1}{(c-2)} |S| (|B| + 2) \tag{6.28}$$

$$\leqslant \frac{d^2}{2(c-2)} (|B| + 2) \overset{\text{eq. (6.25)}}{\leqslant} r, \tag{6.29}$$

since $|S| \leqslant \binom{d}{2} \leqslant \frac{d^2}{2}$ and $|B_F| = |B| + 2$. Hence $|C \cup \{v\}| \leqslant r$. Furthermore, $C \cup \{v\} \subseteq L \setminus A$, so by the fact that (A,B) has the V-matching property, there exists a V-matching F' covering $C \cup \{v\}$ in $G_{A,B}$.

Then there must be some $F \in S$ such that F is a connected component of F' covering v. Let F'' be F' with the component F removed. Then F'' is a V-matching in G_{A_F, B_F} and F'' covers C_F, contradicting the choice of C_F. □

Lemma 6.7 (Covering a Vertex in U). *Let $A \subseteq L$ and $B \subseteq U$ be such that the pair (A,B) has the V-matching property and let v be a vertex in $U \setminus B$ with degree e. If*

$$2(c-2)r \geqslant d^2(|B| + 2e), \tag{6.30}$$

then there exists a V-matching F in $G_{A,B}$ covering v and such that $(A \cup L(F), B \cup U(F))$ has the V-matching property.

A visual hint for the notations used in this proof can be found in Fig. 6.5.

Proof. Fix $v \in U \setminus B$ and let D be $N_G(v) \setminus A$. If $|D| = 0$, then $N_G(v) \subseteq A$, and so we can cover v by taking F to be the V-matching consisting only of the vertex v.

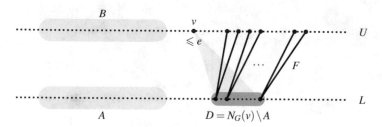

Figure 6.5 Covering a vertex in U via V-matchings

Since $v \in U$, and the graphs $G_\bullet$ from item (a) on p. 72 are allowed in a V-matching, then this is a valid V-matching covering v and clearly $(A \cup L(F), B \cup U(F))$ has the V-matching property.

If $|D| > 0$, by hypothesis $|D| \leqslant e$ and hence, by the cardinality condition on B, see eq. (6.30), we can apply Lemma 6.6 $|D|$ times to obtain a V-matching F in $G_{A,B}$ covering D and such that $(A \cup L(F), B \cup U(F))$ has the V-matching property.

Now, since $N_G(v) \subseteq A \cup L(F)$, it follows that $(A \cup L(F), B \cup U(F) \cup \{v\})$ has the V-matching property. Either v is covered by F, or it is possible to add $\{v\}$ as a new connected component to F while still maintaining the property of being a V-matching in $G_{A,B}$. $\square$

We now have all the preliminary lemmas needed to prove Theorem 6.2 (restated below for the convenience of the reader).

Restated Theorem 6.2 *Let G be a bipartite graph with bipartition $(L(G), U(G))$, r a positive integer and $c > 2$ a real number. Suppose that the following two properties hold:*

1. *G is an (r,c)-bipartite expander;*
2. *the maximum degree of a vertex in $U(G)$ is at most μ.*

Then Cover wins $\mathsf{CovGame}_V(G, \mu)$ with $\mu = \frac{r(c-2)}{2d^2}$ where d is the maximum degree of a vertex in $L(G)$.

Proof. Let $\mathcal{L}$ be the set of all V-matchings F in G such that $(L(F), U(F))$ has the V-matching property, and $|U(F)| \leqslant r(c-2)$.

We claim that Cover can use the V-matchings in $\mathcal{L}$ to win $\mathsf{CovGame}_V(G, \mu)$. By Lemma 6.4 the empty V-matching is in $\mathcal{L}$ and hence $\mathcal{L}$ is non-empty. Moreover, $\mathcal{L}$ is closed under removing connected components by Lemma 6.5. Suppose now that at step $i+1$ of the game Choose picks a vertex v in $G_{L(F_i), U(F_i)}$ and that F_i has strictly fewer than $\mu = \frac{r(c-2)}{2d^2}$ connected components. Then, $(L(F_i), U(F_i))$ satisfies both the cardinality constraints of Lemma 6.6 and Lemma 6.7. Let d_U be the max degree of a vertex in $U(G)$:

$$d^2(U(F_i) + 2d_U) \leqslant d^2(2\mu + 2d_U) \tag{6.31}$$

$$\leqslant d^2(4\mu) \tag{6.32}$$

$$= 2(c-2)r. \tag{6.33}$$

Here eq. (6.31) follows from the fact that $|U(F_i)| \leqslant 2\mu$, and eq. (6.32) follows by the hypothesis that $d_U \leqslant \mu$. The last equality is just the hypothesis on μ.

If v is already covered by F_i we take $F_{i+1} = F_i$. Otherwise, by Lemma 6.6 and Lemma 6.7 applied to $(L(F_i), U(F_i))$, there exists a V-matching F_{i+1} extending F_i by a new connected component covering v such that $(L(F_{i+1}), U(F_{i+1}))$ has the V-matching property. From the previous chain of inequalities, it follows easily that the pair $(L(F_{i+1}), U(F_{i+1}))$ satisfies the cardinality condition $|U(F_{i+1})| \leqslant r(c-2)$. □

6.5.2 A Winning Strategy for the Game on VW-Matchings

The proof of this theorem is analogous to the one of Theorem 6.2, but there are some non-trivial small changes in some crucial lemmas we need.

To simplify the exposition in this subsection we consider a fixed bipartite graph G, an integer r and a real number $c > 1.9$ such that G is an (r,c)-bipartite expander. For brevity let $L = L(G)$, $U = U(G)$ and let each vertex in L have degree at most 3. As in the previous section, given $A \subseteq L$ and $B \subseteq U$, we let $G_{A,B}$ be the subgraph of G induced by $(L \setminus A) \cup (U \setminus B)$.

Definition 6.5 (VW-**Matching Property**). Given two sets $A \subseteq L$ and $B \subseteq U$, we say that the pair (A,B) has the VW-*matching property* if for every $C \subseteq L \setminus A$ with $|C| \leqslant r$, there exists a VW-matching F in $G_{A,B}$ covering C.

Lemma 6.8. *Let* $A \subseteq L$ *and* $B \subseteq U$ *be such that the pair* (A,B) *does not have the* VW-*matching property. Then there exists a set* $C \subseteq L \setminus A$ *with* $(2-c)|C| < |B|$, *such that no* VW-*matching in* $G_{A,B}$ *covers* C.

Proof. Take $C \subseteq L \setminus A$ of minimal size such that no VW-matching in $G_{A,B}$ covers C. We have that $|C| \leqslant r$.

By the minimality of C, every proper subset of C can be covered by a VW-matching and moreover no pair of degree 3 vertices in $L(G)$ have the same set of neighbors, in fact if $A \subseteq L(G)$ has size 2 then $|N(G)| \geqslant 1.9 \cdot 2 = 3.8 > 3$. Then Theorem 6.1 implies that

$$\left| N_{G_{A,B}}(C) \right| < (2c-2)|C|. \tag{6.34}$$

On the other hand, by assumption, G is an (r,c)-bipartite expander, hence

$$c|C| \leqslant N_G(C). \tag{6.35}$$

Therefore we have the following chain of inequalities:

$$c|C| \leqslant N_G(C) \leqslant \left| N_{G_{A,B}}(C) \right| + |B| < (2c-2)|C| + |B| \, . \tag{6.36}$$

Hence $(2-c)|C| < |B|$. $\square$

The previous is the only place where we directly use Theorem 6.1, the version of Hall's theorem for VW-matchings. However, similarly to Sect. 6.5.1, Lemma 6.8 itself plays a crucial role in proving the following lemmas.

Lemma 6.9. *The pair* $(\emptyset, \emptyset)$ *has the* VW-*matching property.*

Proof. For sake of contradiction suppose that $(\emptyset, \emptyset)$ does not have the VW-matching property. Then, by Lemma 6.8, there exists a set $C \subseteq L \setminus A$ that has no VW-matching in $G_{A,B}$ covering C and C has negative size, which is clearly not possible. $\square$

Lemma 6.10 (Component Removal). *Let* $A \subseteq L$ *and* $B \subseteq U$ *be such that the pair* (A, B) *has the* VW-*matching property and*

$$r(2-c) \geqslant |B| \, . \tag{6.37}$$

Then for each VW-*matching* F *contained in the subgraph of* G *induced by* $A \cup B$, $(A \setminus L(F), B \setminus U(F))$ *has the* VW-*matching property.*

A visual hint for the notations used in this proof can be found in Fig. 6.6.

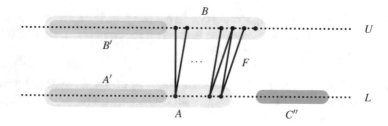

Figure 6.6 Component removal for VW-matchings

Proof. Let $A' = A \setminus L(F)$ and $B' = B \setminus U(F)$ and suppose, for sake of contradiction, that (A', B') does not have the VW-matching property. By Lemma 6.8, it is sufficient to prove that for each set $C \subseteq L \setminus A'$ with $|C| < \frac{1}{2-c}|B'|$, there is a VW-matching in $G_{A',B'}$ covering C. Let $C' = C \cap L(F)$ and $C'' = C \setminus C'$. By construction, F is a VW-matching such that $L(F) \subseteq A$, $U(F) \subseteq B$ and F covers C'. Moreover, we have that

$$|C''| \leqslant |C| < \frac{1}{2-c}|B'| < \frac{1}{2-c}|B| \overset{\text{eq. (6.37)}}{\leqslant} r \, . \tag{6.38}$$

Hence there exists a VW-matching F'' of C'' in $G_{A,B}$, and so $F \cup F''$ is a VW-matching covering C in $G_{A',B'}$. $\square$

Lemma 6.11 (Covering a Vertex in L). *Let $A \subseteq L$ and $B \subseteq U$ be such that the pair (A,B) has the VW-matching property and let d be the maximum degree of a vertex in $U \setminus B$. If*

$$r(2-c) \geqslant 12d(|B|+3),\tag{6.39}$$

then for each vertex v in $L \setminus A$, there is a VW-matching F in $G_{A,B}$ covering v and such that $(A \cup L(F), B \cup U(F))$ has the VW-matching property.

A visual hint for the notations used in this proof can be found in Fig. 6.7.

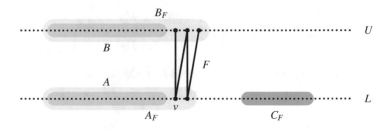

Figure 6.7 Covering a vertex in L via VW-matchings

Proof. Fix $v \in L \setminus A$ and let S be the set of all VW-matchings F in $G_{A,B}$ covering v and such that F is connected.

Since $r \geqslant 1$ and (A,B) has the VW-matching property, we have that S is non-empty. For every $F \in S$, let (A_F, B_F) be the pair $(A \cup L(F), B \cup U(F))$, and suppose for a contradiction that for every $F \in S$, (A_F, B_F) does not have the VW-matching property. By Lemma 6.8 then for every $F \in S$ there is a set $C_F \subseteq L \setminus A_F$ with $|C_F| < \frac{1}{2-c}|B_F|$ and such that there is no VW-matching of C_F in G_{A_F, B_F}. Let $C = \bigcup_{F \in S} C_F$. Then

$$|C| \leqslant \sum_{F \in S} |C_F| < \frac{1}{2-c} \sum_{F \in S} |B_F| \leqslant \frac{1}{2-c} |S|(|B|+3) \leqslant \frac{1}{2-c} 12d(|B|+3) \leqslant r.$$
$$\tag{6.40}$$

Since there are at most three V-matchings covering v and at most $3 \cdot 2 \cdot (d-1) \cdot 2$ W-matchings covering v, we have that $|S| \leqslant 3+3 \cdot 2 \cdot (d-1) \cdot 2 \leqslant 12d$ and moreover $|B_F| \leqslant |B|+3$. Hence, by eq. (6.39), we have that $|C \cup \{v\}| \leqslant r$. Furthermore, $C \cup \{v\} \subseteq L \setminus A$, so by the fact that (A,B) has the VW-matching property, there is a VW-matching F' covering $C \cup \{v\}$ in $G_{A,B}$.

Then there must be some $F \in S$ such that F is a connected component of F'. Let F'' be F' with the component F removed. Then F'' is a VW-matching in G_{A_F, B_F} and F'' covers C_F, contradicting the choice of C_F. □

Lemma 6.12 (Covering a Vertex in U). *Let $A \subseteq L$ and $B \subseteq U$ be such that the pair (A,B) has the VW-matching property and let d be the maximum degree of a vertex in $U \setminus B$. If*

$$r(2-c) \geqslant 12d(|B|+3d),\tag{6.41}$$

then for each vertex v in $U \setminus B$, there is a VW-matching F in $G_{A,B}$ covering v and such that $(A \cup L(F), B \cup U(F))$ has the VW-matching property.

A visual hint for the notations used in this proof can be found in Fig. 6.8.

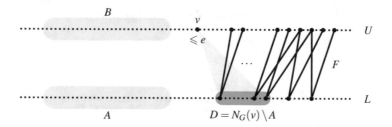

Figure 6.8 Covering a vertex in U via VW-matchings

Proof. Fix $v \in U \setminus B$ and let D be $N_G(v) \setminus A$. If $|D| = 0$, then $N_G(v) \subseteq A$, and so we can cover v by taking F to be the VW-matching consisting only of the vertex v. We have that $v \in U$, and $G_\bullet$ as defined in item (a) on p. 72 is allowed in a VW-matching. Then this is a valid VW-matching covering v and clearly $(A \cup L(F), B \cup U(F))$ has the VW-matching property.

If $|D| > 0$, since by hypothesis $|D| \leqslant d$ and by eq. (6.41), we can then apply Lemma 6.11 $|D|$ times to obtain a VW-matching F in $G_{A,B}$ covering D and such that $(A \cup L(F), B \cup U(F))$ has the VW-matching property.

Now, since $N_G(v) \subseteq A \cup L(F)$, it follows that $(A \cup L(F), B \cup U(F) \cup \{v\})$ has the VW-matching property. Either v is covered by F, or it is possible to add $\{v\}$ as a new connected component to F while still maintaining the property of being a VW-matching in $G_{A,B}$. $\square$

We now have all the preliminary lemmas needed to prove Theorem 6.3 (restated below for the convenience of the reader).

Restated Theorem 6.3 *Let G be a bipartite graph with bipartition $(L(G), U(G))$, r, D be positive integers, and $c > 1.9$ be a real number. For every integer $d \geqslant D$ let S_d be the set of vertices of $U(G)$ with degree bigger than d. Suppose that*

1. *each vertex in $L(G)$ has degree at most 3;*
2. *G is an (r,c)-bipartite expander;*
3. *for every $D_{\max} \geqslant d \geqslant D$,*

$$r(2-c) \geqslant 72d(|S_d| + d) ,\tag{6.10}$$

where $D_{\max}$ is the maximum degree of a vertex in $U(G)$.

Then Cover wins CovGame$_{VW}(G, \mu)$ *with* $\mu = \frac{r(2-c)}{72D}$.

Proof. By the hypothesis on $|S_d|$, for each $D_{\max} \geqslant d \geqslant D$ we can repeatedly apply Lemma 6.12 starting from $(\emptyset, \emptyset)$ to cover vertices in $U(G)$ of degree larger than D. Indeed, by starting by covering vertices of $U(G)$ of maximum degree and proceeding in decreasing order until we have covered the vertices of degree D, we can build a VW-matching F covering S_D such that $(L(F), U(F))$ has the VW-matching property. Moreover, by the choice of S_D, $G_{L(F),U(F)}$ (the subgraph induced by $(L \cup U) \setminus (L(F) \cup U(F))$) has degree at most D. We say that a VW-matching F' is *compatible* with F if each connected component of F' is either a connected component of F or disjoint from all connected components of F.

Let $\mathcal{L}$ be the set of all VW-matchings F' in G compatible with F such that $(L(F) \cup L(F'), U(F) \cup U(F'))$ has the VW-matching property, and moreover such that $|U(F) \cup U(F')| \leqslant \frac{r(2-c)}{2}$. We show now that Cover can use the VW-matchings in $\mathcal{L}$ to win the game $\mathsf{CovGame}_{\mathsf{VW}}(G, \mu)$.

By Lemma 6.9, the empty VW-matching is in $\mathcal{L}$, so this family is non-empty. Moreover, $\mathcal{L}$ is closed under removing connected components by Lemma 6.10. Suppose now that at step $i+1$ of the game Choose picks a vertex v in $G_{L(F),U(F)}$ and that F_i has strictly fewer than $\mu = \frac{r(2-c)}{72D}$ connected components. Then, $(L(F) \cup L(F_i), U(F) \cup U(F_i))$ satisfies the hypotheses of Lemma 6.11 and Lemma 6.12:

$$12D(|U(F) \cup U(F_i)| + 3D) \leqslant 12D(|U(F)| + 3D) + 12D|U(F_i)| \tag{6.42}$$
$$\leqslant 12D(3|S_D| + 3D) + 36D\mu \tag{6.43}$$
$$\leqslant \frac{r(2-c)}{2} + 36D\mu \tag{6.44}$$
$$= \frac{r(2-c)}{2} + 36D\frac{r(2-c)}{72D} \tag{6.45}$$
$$= r(2-c), \tag{6.46}$$

where the inequality (6.43) follows from the fact that $|U(F_i)| \leqslant 3\mu$ and $|U(F)| \leqslant 3|S_D|$, where S_D is the set of vertices in U of degree bigger than D. The inequality (6.44) follows by the hypothesis on the size of S_D.

Hence, if v is covered by F_i we take $F_{i+1} = F_i$. If v is covered by F we take $F_{i+1} = F_i \cup F_v$, where F_v is the connected component of F covering v. Otherwise, by Lemma 6.11 and Lemma 6.12 applied to $(L(F) \cup L(F_i), U(F) \cup U(F_i))$, there exists a VW-matching F_{i+1} extending $F_i \cup F$ by a new connected component covering v such that $(L(F_{i+1}), U(F_{i+1}))$ has the VW-matching property. From the previous chain of inequalities, it follows easily that the pair $(L(F_{i+1}), U(F_{i+1}))$ satisfies the cardinality condition $|U(F) \cup U(F_{i+1})| = |U(F_{i+1})| \leqslant \frac{r(2-c)}{2}$. $\square$

History

The proof of Theorem 6.3 is based on [BBG+17], the one of Theorem 6.2 is based on [BG13, BGT14]. The proofs of the winning strategies are modeled on analogous results in [BGT14, BBG$^+$17] but they are also similar to constructions that can be found in the literature for matchings, for example in [BG03, Ats04]. In particular the definitions of the V-matching property and the VW-matching property (Definition 6.4 and Definition 6.5) are inspired by a similar definition for usual matchings from [Ats04]. Regarding Theorem 6.1, we have that originally it was proven for $c > 1.96$ in [BBG$^+$17], then Susanna Figueiredo de Rezende simplified the argument to show that it holds for $c > 1.8$ (*pers.comm.*). Recently [Rob16] showed that Theorem 6.1 holds for $c \geqslant 5/3$. This is indeed the best possible constraint on c we can get in that theorem. Indeed [BBG$^+$17] showed that Theorem 6.1 becomes false for $c < 5/3$. Anyway, we are not really interested in optimizing the constant c in Theorem 6.1 since, in the applications we show, it will be absorbed in some asymptotic notation.

Chapter 7
Some Graph-Based Formulas

We now show some further applications of the general theorems to prove space lower bounds from Chap. 3 and Chap. 4. That is we see how to prove monomial and resolution total space lower bounds for random k-CNF formulas (Sect. 7.2), the pigeonhole principle over a bipartite graph (Sect. 7.3) and Tseitin formulas (Sect. 7.4).

7.1 From Cover Games to Space Lower Bounds

The way we prove the lower bounds for random k-CNF formulas and the graph pigeonhole principle is to construct some (w,I)-BG families from the winning strategies we saw in Chap. 6. This is the informal content of Lemma 7.1 and it is what we prove in this section. We follow the notations used in Sect. 4.5.

Let Y be a set of variables and $M = \{m_j\}_{j \in J}$ be a set of monomials in the ring of polynomials $\mathbb{F}[Y]$. Let G_M be the *adjacency graph* of M, that is the bipartite graph with lower part $L(G_M) = J$, upper part $U(G_M) = Y$ and there is an edge $\{j,y\}$ in G_M if and only if $y \in \text{var}(m_j)$. This definition generalizes immediately considering families of Boolean assignments instead of variables and in particular this will be helpful in Sect. 7.3.

Definition 7.1 ($\mathcal{A}$-Adjacency Graph). Given a field $\mathbb{F}$, a set of variables Y, collection of families of Boolean assignments $\mathcal{A} = \{A_1, \ldots, A_s\}$ and $M = \{m_j\}_{j \in J}$ a set of monomials in $\mathbb{F}[Y]$. The $\mathcal{A}$-*adjacency graph* of $M = \{m_j\}_{j \in J}$ is the bipartite graph $G_M^{\mathcal{A}}$ with lower part $L(G_M^{\mathcal{A}}) = J$, upper part $U(G_M^{\mathcal{A}}) = \mathcal{A}$ and (j, A_ℓ) is an edge in $G_M^{\mathcal{A}}$ if and only if $\text{var}(m_j) \cap \text{dom}(A_\ell) \neq \emptyset$.

We then have that winning strategies for the cover game on this particular kind of bipartite graphs generate space lower bounds, via the existence of suitable (μ, I)-BG families, see Definition 4.5.

© Springer International Publishing AG, part of Springer Nature 2017
I. Bonacina, *Space in Weak Propositional Proof Systems*,
https://doi.org/10.1007/978-3-319-73453-8_7

Lemma 7.1. *Let* $\mathbb{F}$ *be a field, Y a set of variables, $M = \{m_j\}_{j \in J}$ a set of monomials and I a proper ideal in the ring $\mathbb{F}[Y]$. Suppose we have a collection $\mathcal{A}$ of flippable families of Boolean assignments $A_1, \ldots, A_s$ in the variables Y that are I-consistent and domain-disjoint. If Cover wins* $\mathsf{CovGame}_{\mathcal{C}}(G_M^A, \mu)$ *with $\mathcal{C}$ a collection of trees with no leaves in $L(G_M^A)$, then for every set of polynomials P generating I there exists a (μ, I)-BG family $\mathcal{F}$ for $M \cup P$. Moreover if for each polynomial $p \in I$ there exists an $A_i \in \mathcal{A}$ such that $\mathrm{var}(p) \subseteq \mathrm{dom}(A_i)$ then $\mathcal{F}$ is a $(\mu, 0)$-BG family for $M \cup P$.*

In particular, by Theorem 4.2, this immediately implies that for every I-semantic polynomial calculus refutation π of $M \cup P$ over $\mathbb{F}[Y]$

$$\mathrm{MSp}(\pi) \geqslant \lfloor \mu/4 \rfloor . \tag{7.1}$$

Proof. Given F a $\mathcal{C}$-matching in G_M^A with connected components $F_1, \ldots, F_c$ first we claim that there exist flippable product-families of Boolean assignments H_{F_i} such that

1. $H_{F_i} \vDash_0 \{m_j \, : \, j \in L(F_i)\}$,
2. $H_{F_i} \subseteq \bigotimes_{A_\ell \in U(F_i)} A_\ell$, and
3. $\|H_{F_i}\| = 1$.

Then we define $H_F = H_{F_1} \otimes \cdots \otimes H_{F_c}$. Suppose for a moment that this holds, then we can take as family of flippable Boolean assignments $\mathcal{F}$ the family of all H_F with F appearing in a given winning strategy for Cover in $\mathsf{CovGame}_{\mathcal{C}}(G_M^A, \mu)$. We claim that $\mathcal{F}$ is a (μ, I)-BG family for $M \cup P$.

The *I-consistency* property of $\mathcal{F}$ follows immediately from the I-consistency of the families A_i. The *restriction* property is immediate too. To prove the *extension* property suppose we have some $H_F \in \mathcal{F}$ with $\|H_F\| < \mu$ and some p in $M \cup P$. Consider first the case when $p = m_j \in M$. If F already covers j we have nothing to extend, $H_F \vDash_0 m_j$. If not, thanks to the winning strategy of Cover, there exists some F_j covering j in G_M^A disjoint from F. Then by construction $H' = H_F \otimes H_j \in \mathcal{F}$, $H' \sqsupseteq H_F$ and $H' \vDash_0 m_j$. In the case $p \in P$, we do nothing since H_F is I-consistent and hence already by definition $H_F \vDash_I p$.

Let's check then that if for each $p \in I$ there exists an $A_i \in \mathcal{A}$ such that $\mathrm{var}(p) \subseteq \mathrm{dom}(A_i)$, then $\mathcal{F}$ is a $(\mu, 0)$-BG family for $M \cup P$. We then just have to check the *extension* property when $p \in I$. In this case, by hypothesis, there exists some $A_i \in \mathcal{A}$ such that $\mathrm{var}(p) \subseteq \mathrm{dom}(A_i)$. Similarly to before, by the winning strategy of Cover, then there exists some H' in $\mathcal{F}$ such that $\mathrm{dom}(A_i) \subseteq \mathrm{dom}(H')$, hence, for each Boolean assignment $\alpha \in H'$, $p \restriction_\alpha \in \mathbb{F}$. But since H' is I-consistent and $p \in I$ then $p \restriction_\alpha \in I$. Since I is a proper ideal $p \restriction_\alpha \in I \cap \mathbb{F} = 0$ and hence $\alpha \vDash_0 p$.

It only remains to check the claim at the beginning of this proof, the fact that from $\mathcal{C}$-matchings we can construct flippable product-families. Suppose then F is a $\mathcal{C}$-matching in G_M^A and F consists of a single connected component. We prove the desired properties by induction on the size of the tree F. We prove the additional property that for every $A_\ell \in U(F)$ and every $\alpha \in A_\ell$ there exists $\beta \in H_F$ such that $\beta \supseteq \alpha$.

If F doesn't cover any $j \in L(G_M^A)$ then we can surely take $H_F = \bigotimes_{A_\ell \in U(F)} A_\ell$. So consider the minimal possible *non-trivial* tree with all its leaves in $U(G_M^A)$. It is a tree isomorphic to G_V (see item (b) on p. 72) and without loss of generality we can assume that F is as in Fig. 7.1. Then, analogously to what we did in Example 4.1, let

$$H_F = (O_{m_j,i} \otimes A_\ell) \cup (A_i \otimes O_{m_j,\ell}), \tag{7.2}$$

where $O_{m_j,k} = \{\alpha \in A_k : \alpha \vDash_0 m_j\}$. Since both A_i and A_ℓ are flippable then H_F is non-empty and flippable. Moreover, by construction, $H_F \vDash_0 m_j$ and $\|H_F\| = 1$.

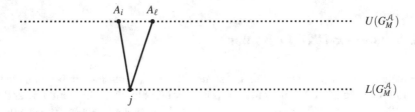

Figure 7.1 From $\mathcal{C}$-matchings to flippable products: a minimal example

Consider now a non-minimal tree F. If the tree F is a star, that is all vertices in $U(F)$ are leaves, then we just take two such leaves and reason as before. Otherwise there exists a vertex A_s in $U(G_M)$ that is not a leaf of F and such that F is the union of two trees F' and F'' whose vertices intersect only on A_s, see Fig. 7.2.

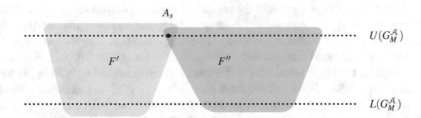

Figure 7.2 From $\mathcal{C}$-matchings to flippable products: inductive step

By the inductive hypothesis, there exists a flippable family $H_{F'}$ such that $\|H_{F'}\| = 1$, $H_{F'} \subseteq \bigotimes_{A_\ell \in U(F')} A_\ell$ and $H_{F'} \vDash_0 \{m_j : j \in L(F')\}$. Similarly we have the same properties for $H_{F''}$. Let H_F be the set of Boolean assignments obtained by 'gluing' together compatible Boolean assignments from $H_{F'}$ and $H_{F''}$. More precisely H_F is the set of all Boolean assignments β in $\bigotimes_{A_\ell \in U(F)} A_\ell$ such that $\beta = \alpha' \cup \alpha''$ with $\alpha' \in H_{F'}$ and $\alpha'' \in H_{F''}$ that both extend some $\alpha \in A_s$:

$$\exists \alpha \in A_s \, \exists \alpha' \in H_{F'} \, \exists \alpha'' \in H_{F''}, \alpha' \supseteq \alpha, \; \alpha'' \supseteq \alpha \text{ and } \beta = \alpha' \cup \alpha''. \tag{7.3}$$

Now let $i \in U(F) = U(F') \cup U(F'')$. We prove that each Boolean assignment in A_i can be extended to a Boolean assignment in H_F, hence the inductive hypothesis will be proved and in particular H_F is flippable. Without loss of generality let $i \in U(F')$ and let $\delta \in A_i$. By the inductive property on $H_{F'}$ there exists some $\alpha' \in H_{F'}$ such that $\alpha' \supseteq \delta$. By the inductive hypothesis on $H_{F'}$, there exists $\alpha \in A_s$ such that $\alpha = \alpha'\lceil_{\text{dom}(A_s)}$. By the inductive property on $H_{F''}$ there exists $\alpha'' \in H_{F''}$ such that $\alpha'' \supseteq \alpha$. Then, by construction, $\beta = \alpha' \cup \alpha'' \in H_F$ and clearly $\beta \supseteq \delta$. Notice that the property that the A_js are domain-disjoint guarantees that $\text{dom}(\alpha') \cap \text{dom}(\alpha'') = A_s$, which is ultimately the property guaranteeing that we can glue together in a consistent way the families given by the inductive hypothesis. □

7.2 Random k-CNF Formulas

The main results of this section are an asymptotically optimal monomial space lower bound for random k-CNF formulas in polynomial calculus, Theorem 7.1, and an asymptotically optimal total space lower bound for those formulas in resolution, Theorem 7.2. But first things first, let's first recall the formal definition of random k-CNF formulas and some properties of those formulas, to give a bit of context.

Definition 7.2 ((n, k, Δ)-Random CNF Formulas). Given a positive integer k and a positive real number Δ, an (n, k, Δ)-*random CNF* is a k-CNF formula with n variables and Δn clauses picked uniformly at random from the set of all CNF formulas in the variables $\{x_1, \ldots, x_n\}$ that consist of exactly Δn clauses, each clause containing exactly k literals and no variable appearing twice in a clause.

A fundamental conjecture about the (n, k, Δ)-random CNF formulas says that there exists a constant θ_k, the *satisfiability threshold*, such that if $\Delta > \theta_k$ then an (n, k, Δ)-random CNF formula is almost surely unsatisfiable, while if $\Delta < \theta_k$ then an (n, k, Δ)-random CNF formula is almost surely satisfiable, see for instance [CS88, KKKS98, FS96, BFU93, CF90]. It is known that for each n there exists a threshold $\theta_k(n)$ with this property [Fri98]. For $k = 2$ we have that $\theta_2(n) = 1$ [CR92, Goe96, dlV01]. In general for $k \geq 3$ and for each n, $\theta_k(n)$ is bounded between two constants that are independent of n, e.g., $3.003 \leq \theta_3(n) \leq 4.598$ [KKKS98, FS96]. Moreover it was recently shown that for large k there exists an explicit constant γ_k not depending on n such that $\theta_k(n) = \gamma_k$ [DSS15]. It is believed that (n, k, Δ)-random CNF formulas with Δ close to the satisfiability threshold θ_k are the ones for which it is most computationally hard to show they are unsatisfiable, see for example [CKT91].

Regarding the upper bounds, it is easy to see that the $(n, 2, \Delta)$-random CNF formulas are easy for resolution since the easy well-known polynomial-time algorithm to solve 2-SAT can be formalized to give polynomial-size resolution refutations of any unsatisfiable 2-CNF formula. Regarding the $(n, 3, \Delta)$-random CNF formulas, [BKPS98] showed that for any $(n, 3, \Delta)$-random CNF formula with $\Delta > \theta_3$, with high probability there exists a resolution refutation of size at most $2^{O(n/\Delta)}$, which is a function of polynomial growth when $\Delta \geq n/\log n$.

Regarding the lower bounds, [CS88] showed that every (n,k,Δ)-random CNF formula, with $k \geqslant 3$ and Δ a constant such that $\Delta > \theta_k$, is hard for resolution to refute. That is with high probability every resolution refutation of those formulas has size at least $2^{\Omega(n)}$. The importance of this result relies on the fact that it proves that resolution is a very weak propositional proof system, in the sense that almost all 3-CNF formulas require exponential-size resolution refutations. Since this seminal result, the hardness of (n,k,Δ)-random CNF formulas has been deeply investigated; in particular this lower bound was improved and simplified by [BP96], improved to $\Delta = o(n^{1/4})$ by [BKPS02] and simplified using the size-width inequality, see eq. (2.10), by [BW01]. All these results, as well as the ones we show in this section, hold for $k \geqslant 3$. Moreover, the (n,k,Δ)-random CNF formulas have been shown to be hard to refute also in polynomial calculus, see [BI10, AR01], and for resolution over k-DNF formulas, see [Ale11]. Recently it has been shown that $(n, O(\log n), \Delta)$-random CNF formulas for large enough Δ are exponentially hard to refute for cutting planes [HP17, FPPR17].

It is not known whether (n,k,Δ)-random CNF formulas with k constant are hard for cutting planes or bounded-depth Frege, although this is usually conjectured to be the case.

With respect to the space complexity of (n,k,Δ)-random CNF formulas, they have been shown to require large clause space in resolution. More precisely given an (n,k,Δ)-random CNF formula F with $\Delta > \theta_k$ with high probability any resolution refutation π of F has

$$\mathrm{CSp}(\pi) = \Omega(n/\Delta^{1+\varepsilon}), \tag{7.4}$$

see [BG03]. On the other hand [Zit02] showed an upper bound on their clause space: there are resolution refutations π' of F such that

$$\mathrm{CSp}(\pi') = O(n\Delta^{-1/(k-2)}). \tag{7.5}$$

Regarding monomial space and total space (in resolution) we have the following two asymptotically optimal results, conjectured to be true in several works in the literature, e.g., [BS01, ABRW02, FLN$^+$15].

Theorem 7.1 (Monomial Space Lower Bound). *Let $\mathbb{F}$ be a field, $k \geqslant 3$, $\Delta > 1$ a constant and F an (n,k,Δ)-random CNF formula over the variables $X = \{x_1,\ldots,x_n\}$. Then, for large n, with high probability, for every I-semantic polynomial calculus refutation π of the polynomial encoding of F in $\mathbb{F}[X \cup \overline{X}]$,*

$$\mathrm{MSp}(\pi) = \Omega(n), \tag{7.6}$$

where I is the ideal in $\mathbb{F}[X \cup \overline{X}]$ generated by the Boolean axioms.

Proof. Let $F = \bigwedge_{j \in J} C_j$ be an (n,k,Δ)-random CNF formula and let G_F be its clauses-variables adjacency graph. That is the bipartite graph with bipartition $(L(G_F), U(G_F))$ with $L(G_F) = J$ and $U(G_F) = \{x_1,\ldots,x_n\}$, and $\{j,x_i\}$ is an edge in G_F if $x_i \in \mathrm{var}(C_j)$. Then G_F is an (n,k,Δ)-random bipartite graph, see Sect. 6.4. Now, if $k > 3$ then, by Theorem 6.5, for large n, with high probability there exists a constant γ such that Cover has a winning strategy for $\mathsf{CovGame}_V(G_F, \gamma n)$. Similarly, if $k = 3$, by Theorem 6.5, for large n, with high probability there exists a constant γ such that Cover has a winning strategy for $\mathsf{CovGame}_{VW}(G_F, \gamma n)$. Consider the following collection of families of flippable Boolean assignments $\mathcal{A} = \{A_1, \ldots, A_n\}$ with $A_i = \{\alpha_i, \alpha_i'\}$ where $\mathrm{dom}(\alpha_i) = \mathrm{dom}(\alpha_i') = \{x_i, \bar{x}_i\}$ and

$$\alpha_i(x_i) = 1 - \alpha_i(\bar{x}_i) = \alpha_i'(\bar{x}_i) = 1 - \alpha_i'(x_i) = 0. \tag{7.7}$$

The polynomial encoding of F in $\mathbb{F}[X \cup \overline{X}]$ is the union of a set of monomials $M = \{m_j : j \in J\}$ encoding the clauses of F and the set of Boolean axioms B. We have that G_M^A is equivalent to G_F from the point of view of its graph structure. So we can apply Lemma 7.1 to obtain that there exists a $(\gamma n, I)$-BG family for $M \cup B$ where I is the ideal generated by B. Then the monomial space lower bound follows from Theorem 4.2. $\square$

Notice that in the previous proof, from the properties of a random bipartite graph, we are just using the fact that it satisfies the hypotheses of Theorem 6.2 and Theorem 6.3. That is the same result will hold for any CNF formula F_n with its clauses-variables adjacency graph satisfying the hypotheses of Theorem 6.2 and Theorem 6.3.

Theorem 7.2 (Total Space Lower Bound). *Let $k \geqslant 3$, $\Delta > 1$ a constant and F an (n, k, Δ)-random CNF over the variables $X = \{x_1, \ldots, x_n\}$. Then, for large n, with high probability, for every resolution refutation π of F,*

$$\mathrm{TSp}(\pi) = \Omega\left(n^2\right). \tag{7.8}$$

Proof. Inspecting the proof of the previous theorem, it is immediate to see that the $(\gamma n, I)$-BG family $\mathcal{F}$ for $M \cup B$ is indeed a $(\gamma n, 0)$-BG family for $M \cup B$. Indeed for every $p \in B$ there exists some A_i such that $\mathrm{var}(p) \subseteq A_i$ and then we can apply Lemma 7.1. Hence we can apply Proposition 4.5 and immediately obtain the desired total space lower bound on resolution. $\square$

Alternatively the total space lower bound also follows from the fact that for large n, with high probability, there exists a constant $\gamma > 0$ such that every resolution refutation π of F requires width at least γn, see [CS88, BW01]. Hence this total space lower bound on resolution follows also from Theorem 3.6.

As an immediate consequence of this total space lower bound on resolution we have an optimal separation between *semantic* resolution (see Sect. 3.1) and resolution. Indeed for Δ a large enough constant, (n, k, Δ)-random CNF formulas have semantic resolution refutations of total space $O(n)$ but the previous theorem tells us that every usual resolution refutation of those formulas must use quadratic total space. This observation completely answers [ABRW02, Open question 4] for resolution.

Despite this separation we can strengthen Theorem 7.2 as follow.

Theorem 7.3 (dn-Semantic Total Space Lower Bound). *Let $k \geqslant 3$, $\Delta > 1$ a constant and F an (n, k, Δ)-random CNF over the variables $X = \{x_1, \ldots, x_n\}$. Then,* for large n there exists a constant d such that, with high probability, for every dn-semantic resolution refutation π of F,

$$\text{TSp}(\pi) = \Omega\left(n^2\right) . \tag{7.9}$$

Proof. This follows immediately from the linear width resolution lower bound for random k-CNF formulas and Theorem 3.8.

7.3 Pigeonhole Principles over Graphs

In this section we continue the investigation of the pigeonhole principles, so in some sense this is a continuation of Sect. 5.1.

Let G be a bipartite graph with bipartition $(L(G), U(G))$ with $|L(G)| > |U(G)|$. We think of $L(G)$ as a set of *pigeons* and $U(G)$ as a set of *holes*. The *graph* pigeonhole principle over the graph G, G-PHP, is an unsatisfiable CNF formula in the variables

$$X = \{x_{uv} : \{u,v\} \in E(G)\} . \tag{7.10}$$

It asserts that the variables describe a map, given by a subset of the edges of G, in which each pigeon gets mapped to at least one hole but no hole receives two pigeons or more. Formally, it is a conjunction of all the following clauses:

1. for each distinct pair of variables $x_{uv}, x_{u'v} \in X$, $\neg x_{uv} \vee \neg x_{u'v}$ (**Hole Axioms**);
2. for each $u \in L(G)$, $\bigvee\{x_{uv} : x_{uv} \in X\}$ (**Pigeon Axioms**).

Notice that if $\max_{v \in L(G)} \deg_G(v) \leqslant d$ then G-PHP is a d-CNF formula and, since $|L(G)| > |U(G)|$, it is an unsatisfiable CNF formula. The graph pigeonhole principle is then a generalization of the standard pigeonhole principle we saw in Sect. 5.1: indeed PHP_n^m is the graph pigeonhole principle $K_{m,n}$-PHP, where $K_{m,n}$ is the complete bipartite graph between a set of vertices of size m and a (disjoint) set of vertices of size n.

Recall that the encoding of G-PHP as a set of polynomials $P_{G\text{-PHP}}$ in $\mathbb{F}[X \cup \overline{X}]$ is completely analogous to the one we saw for the pigeonhole principle, see Sect. 5.1, that is the following:

$$P_{G\text{-PHP}} = \left\{x_{uv}x_{u'v} : x_{uv}, x_{u'v} \in X \text{ and } u \neq u'\right\}$$

$$\cup \left\{ \prod_{v \,:\, x_{uv} \in X} \bar{x}_{uv} : u \in L(G) \right\}$$

$$\cup \left\{x_{uv}^2 - x_{uv}, \, x_{uv} + \bar{x}_{uv} - 1 : \{u,v\} \in E(G)\right\} . \tag{7.11}$$

Matching principles over graphs have been well studied in proof complexity. The interested reader may, for example, look at [BW01, BG03] or [Juk12, Section 18.1]. Let's just proceed straight to the point: to prove monomial and total space lower bound for those formulas.

Theorem 7.4 (Monomial Space Lower Bound). *Let* $\mathbb{F}$ *be a field,* $d \geqslant 3$, $\Delta > 1$ *and* G *an* (n, d, Δ)*-random bipartite graph. Then, for large n, with high probability, for every I-semantic polynomial calculus refutation* π *of the polynomial encoding of* G*-PHP in* $\mathbb{F}[X \cup \overline{X}]$,

$$\mathrm{MSp}(\pi) = \Omega(n) , \tag{7.12}$$

where I is the ideal in $\mathbb{F}[X \cup \overline{X}]$ *generated by polynomial encodings of the hole axioms and the Boolean axioms of G-PHP, that is the ideal generated by the set P of polynomials*

$$P = \left\{ x_{uv} x_{u'v} \ : \ x_{uv}, x_{u'v} \in X \ and \ u \neq u' \right\} \cup \left\{ x_{uv}^2 - x_{uv}, \ x_{uv} + \bar{x}_{uv} - 1 \ : \ x_{uv} \in X \right\} . \tag{7.13}$$

Proof. Since G is an (n, d, Δ)-random bipartite graph, then, if $d > 3$, by Theorem 6.5, for large n, with high probability there exists a constant γ such that Cover has a winning strategy for $\mathsf{CovGame}_V(G, \gamma n)$. Similarly, if $d = 3$, by Theorem 6.5, for large n, with high probability there exists a constant γ such that Cover has a winning strategy for $\mathsf{CovGame}_{VW}(G, \gamma n)$. Let $M = \left\{ \prod_{v \ : \ x_{uv} \in X} \bar{x}_{uv} \ : \ u \in L(G) \right\}$, then the polynomial encoding of G-PHP is $M \cup P$. Then, as done in the proof of Theorem 7.1, it is sufficient to construct a collection $\mathcal{A}$ of flippable product-families satisfying the hypotheses of Lemma 7.1, such that the bipartite graph $G_M^{\mathcal{A}}$ is isomorphic to G.

Let $\mathcal{A} = \{A_v \ : \ v \in U(G)\}$ where $A_v = \{\alpha_{uv} \ : \ \{u, v\} \in E(G)\}$ and α_{uv} is the Boolean assignment with domain $\{x_{u'v}, \bar{x}_{u'v} \ : \ \{u', v\} \in E(G)\}$ and such that

$$\alpha_{uv}(x_{u'v}) = 1 - \alpha_{uv}(\bar{x}_{u'v}) = \begin{cases} 1 & \text{if } u' = u , \\ 0 & \text{if } u' \neq u . \end{cases} \tag{7.14}$$

Clearly we have that A_v is flippable; $\mathrm{dom}(A_v) = X_v$ and hence, if $v \neq v'$ then $\mathrm{dom}(A_v)$ and $\mathrm{dom}(A_{v'})$ are disjoint. Moreover A_v is I-consistent, where I is the ideal generated by P. Moreover, an edge $\{u, v\}$ is in $E(G)$ if and only if

$$\mathrm{var}\left(\prod_{v \ : \ x_{uv} \in X} \bar{x}_{uv} \right) \cap \mathrm{dom}(A_v) \neq \emptyset , \tag{7.15}$$

hence G and $G_M^{\mathcal{A}}$ are isomorphic. Then Lemma 7.1 implies that there is a non-empty $(\gamma n, I)$-BG family for $M \cup P$ and Theorem 4.2 then implies the monomial space lower bound. $\square$

Notice that in the previous proof, from the properties of a random bipartite graph, we are just using the fact that it satisfies the hypotheses of Theorem 6.2 and Theorem 6.3. That is the same result will hold for any graph G satisfying the hypotheses of Theorem 6.2 and Theorem 6.3.

Theorem 7.5 (Total Space Lower Bound). *Let* $d \geqslant 3$, $\Delta > 1$ *and* G *an* (n, d, Δ)*- random bipartite graph. Then, for large n, with high probability, for every resolution refutation* π *of G-PHP,*

$$\mathrm{TSp}(\pi) = \Omega(n^2) . \tag{7.16}$$

Proof. In the proof of Theorem 7.4, the $(\gamma n, I)$-BG family for $M \cup P$ is also a $(\gamma n, 0)$-BG family for $M \cup P$. Indeed for each $p \in I$ there exists some A_v such that $\mathrm{var}(p) \subseteq A_v$ and then we can apply Lemma 7.1. Then the total space lower bound follows from Proposition 4.5. $\square$

Alternatively, as for Theorem 7.2, we could have proven that each resolution refutation of G-PHP requires width at least $\gamma' n$ and then the total space follows from Theorem 3.6. This result then clearly generalizes to dn-semantic resolution refutations of G-PHP.

Theorem 7.6 (dn-**Semantic Total Space Lower Bound**)**.** *Let* $d \geqslant 3$, $\Delta > 1$ *and* G *an* (n, d, Δ)-*random bipartite graph. Then, for large n there exists a constant d such that, with high probability, for every dn-semantic resolution refutation* π *of* G-PHP,

$$\mathrm{TSp}(\pi) = \Omega\left(n^2\right) . \tag{7.17}$$

7.4 Tseitin Formulas

We conclude this chapter by recalling some results on Tseitin formulas. *Tseitin formulas* are propositional formulas encoding the fact that the total degree in any graph is an even number. More precisely, given any graph G with vertex-set V and edge-set E, we have Boolean variables $X = \{x_e : e \in E\}$. Then the Tseitin formula encodes as a CNF formula the fact that

$$\sum_{v \in V} \sum_{e \ni v} x_e \equiv 1 \pmod{2}, \tag{7.18}$$

which is clearly a contradiction since in the previous sum each variable x_e is counted twice: once for one endpoint of e and once for the other. Usually in proof complexity a weaker version of eq. (7.18) is considered where to each vertex $v \in V$ is additionally assigned a weight $w(v) \in \{0, 1\}$ and it is required that for each $v \in V$, $\sum_{e \ni v} x_e \equiv w(v)$ (mod 2). Then if $\sum_{v \in V} w(v) \equiv 1 \pmod{2}$ this task is in some sense even "more impossible" than eq. (7.18). Then the *Tseitin formula* $\mathsf{Tseitin}(G, w)$ is the natural encoding as a CNF formula of this principle. More formally, given a weight function $w : V \to \{0, 1\}$ such that

$$\sum_{v \in V} w(v) \equiv 1 \pmod{2}, \tag{7.19}$$

for every vertex $v \in V$ consider the CNF formula $\mathsf{PARITY}_{v,w}$ naturally encoding the fact that

$$\sum_{e \ni v} x_e \equiv w(v) \pmod{2}. \tag{7.20}$$

Then the *Tseitin formula* is $\mathsf{Tseitin}(G, w) = \bigwedge_{v \in V} \mathsf{PARITY}_{v,w}$. As argued before $\mathsf{Tseitin}(G, w)$ is an unsatisfiable CNF formula. Moreover, if G has maximum degree

d then $\mathsf{Tseitin}(G,w)$ is a d-CNF formula over at most $dn/2$ variables and with at most $n2^{d-1}$ clauses. Moreover, if w is *odd-weight*, that is if w satisfies eq. (7.20), then $\mathsf{Tseitin}(G,w)$ is unsatisfiable (and the other implication is also true) [Urq95].

Tseitin formulas were originally used to give the first super-polynomial lower bounds on refutation size for regular resolution [Tse83]. Then this result was improved to an exponential lower bound on the size of resolution refutations [Urq87, Sch97]. Since then Tseitin formulas became one of the standard tools used in proof complexity to prove lower bounds and trade-offs, for example they have been investigated regarding the resolution width [BW01], clause space [ET01] and size-space trade-offs in polynomial calculus [BNT13]. Recently Håstad proved that Tseitin formulas over square grid graphs with n vertices require super-polynomially long refutation in depth-d Frege where $d = o(\log n / \log\log n)$ (*pers. comm.*).

It turns out that many properties of the proof complexity of Tseitin formulas $\mathsf{Tseitin}(G,w)$ can be captured by the *connectivity expansion* of G.

Definition 7.3 (Connectivity Expansion). Let $G = (V,E)$ be a finite connected graph. The *connectivity expansion* of G, $e(G)$, is

$$e(G) = \min_{V' \subseteq V} \left\{ \{v',v''\} \in E \; : \; v' \in V', \, v'' \in V \setminus V' \text{ and } \frac{|V'|}{|V|} \in \left[\frac{1}{3}, \frac{2}{3}\right] \right\} . \quad (7.21)$$

Then we say that G is a *connectivity expander graph* if $e(G) = \Omega\,(|V|)$, for instance random d-regular graphs with high probability are connectivity expanders [Urq87].

Theorem 7.7 ([BW01, Theorem 4.4]). *Given a connected graph $G = (V,E)$ and an odd-weight function w on V, then every resolution refutation of $\mathsf{Tseitin}(G,w)$ has width at least $e(G)$.*

From this result and eq. (2.10), it follows immediately that under the hypotheses of Theorem 7.7, every resolution refutation π of $\mathsf{Tseitin}(G,w)$ has size

$$S(\pi) = 2^{\Omega\left(\frac{(e(G)-d)^2}{m}\right)}, \quad (7.22)$$

where d is the maximum degree of G and m is the number of edges in G. Similarly a clause space lower bound can be obtained from Theorem 3.2: for every resolution refutation π of $\mathsf{Tseitin}(G,w)$

$$CSp(\pi) > e(G) - d . \quad (7.23)$$

Concerning total space lower bounds in resolution, we have the following result that answers the open question [ABRW02, Open question 2].

Theorem 7.8. *Let $G = (V,E)$ be a connected d-regular graph and w an odd-weight function over V, then for every resolution refutation π of $\mathsf{Tseitin}(G,w)$*

$$TSp(\pi) \geqslant \lfloor (e(G) - d - 4)/4 \rfloor^2 . \quad (7.24)$$

In particular if G is a 3-regular expander graph over n vertices then for every resolution refutation π of Tseitin(G, w)

$$\mathrm{TSp}(\pi) = \Omega\left(n^2\right),\tag{7.25}$$

and the same conclusion holds if π is a cn-semantic resolution refutation for some small enough constant c.

Proof. It follows immediately from Theorem 7.7, Theorem 3.8 and Theorem 3.6. □

Regarding the monomial space in polynomial calculus the picture is more complex. Relying on a preliminary version of Theorem 4.2, [FLM$^+$13] showed the following result about the xorification of CNF formulas.

The *xorification* of a CNF formula F is a new CNF formula $F[\oplus]$ obtained by replacing each occurrence of a variable x_i in F with the XOR of two new variables $x_i' \oplus x_i''$ and then expanding everything as a CNF formula using the definition of XOR and the De Morgan rules. (We will see more general xorifications in Part III.)

Theorem 7.9 ([FLM$^+$13]). *Given a k-CNF formula F over a set of variables X and a field $\mathbb{F}$. If every resolution refutation of F requires width at least W, then for every semantic polynomial calculus refutation π of the polynomial encoding of $F[\oplus]$ in $\mathbb{F}[X \cup \overline{X}]$*

$$\mathrm{MSp}(\pi) \geqslant \frac{1}{4}(W - k + 1).\tag{7.26}$$

In particular if $G = (V, E)$ is a d-regular graph with double edges[1] and w any odd-weight function over V, then for every semantic polynomial calculus refutation π of Tseitin(G, w)

$$\mathrm{MSp}(\pi) = \Omega(e(G) - d).\tag{7.27}$$

Proof (sketch). Eq. 7.26 can be proven essentially by showing that from an r-AD family for F we can construct a suitable $(r', 0)$-BG family for $P_{F[\oplus]}$, And then using Theorem 2.5 and Theorem 4.2. Then eq. (7.27) follows since Tseitin$(G, w)[\oplus]$ is equivalent to Tseitin(G', w) where G' is a multigraph over the vertex-set of G obtained by doubling the multiplicity of each edge of G.   □

As an application of (a preliminary version of) Theorem 4.2, [FLM$^+$13] showed also the following monomial space lower bound for random d-regular graphs with $d \geqslant 4$.

Theorem 7.10 ([FLM$^+$13]). *Let $G = (V, E)$ be a random d-regular graph on n vertices, where $d \geqslant 4$, and let $\mathbb{F}$ be a field. Then, with high probability, for every odd-weight function w on V and every semantic polynomial calculus refutation π of the polynomial encoding of* Tseitin(G, w) *over $\mathbb{F}[X \cup \overline{X}]$*

$$\mathrm{MSp}(\pi) = \Omega\left(\sqrt{n}\right).   \square\tag{7.28}$$

---

[1] That is each edge has multiplicity 2.

Notice that unlike the results we saw on random k-CNF formulas and G-PHP, this result relies more deeply on the fact that G is random.

Over fields of characteristic 2, it is known that the polynomial encoding of Tseitin formulas has polynomial-size refutations in polynomial calculus, essentially mimicking Gaussian elimination. On the other hand, the previous monomial space lower bound does not depend on the characteristic of the ground field. That is, for instance, despite Tseitin formulas over $\mathbb{F}_2$ having short proofs, such refutations still require reasonably large monomial space.

7.5 Open Problems

Question 7.1. Let F be an (n, k, Δ)-random CNF formula. Is it true that for every polynomial calculus refutation π of F

$$\mathrm{TSp}(\pi) = \Omega\left(n^2\right) \text{ ?} \tag{7.29}$$

More generally, all the open questions on total space in polynomial calculus from [ABRW02] are still open. Moreover we have the following open questions.

Question 7.2. Given any constant $c > 1$, a constant γ and an unsatisfiable CNF formula F in n variables such that the clauses-variables adjacency graph of F is a $(\gamma n, c)$-bipartite expander, is this expansion property enough to have that for every polynomial calculus refutation π of F

$$\mathrm{MSp}(\pi) = \Omega\left(n\right) \text{ ?} \tag{7.30}$$

This is the case for the clause space in resolution [BG03] and indeed we suspect that the same happens for the monomial space. Notice that we proved it for $c > 1.9$ and some additional assumptions on the adjacency graph. This is a consequence of Lemma 7.1, Theorem 6.3 and Theorem 4.2.

Question 7.3. Is it true that for every 3-regular graph G that is a connectivity expander with n vertices and w an odd-weight function, for every polynomial calculus refutation π of $\mathsf{Tseitin}(G, w)$,

$$\mathrm{MSp}(\pi) = \Omega\left(n\right) \text{ ?} \tag{7.31}$$

That is we are asking whether we can improve the monomial space lower bound in Theorem 7.10 and at the same time weaken its hypothesis.

History

The monomial space result for random k-CNF formulas for $k \geqslant 4$, Theorem 7.1, and the monomial space for matching principles over left degree-k bipartite expanders,

Theorem 7.4, were proven in [BG13, BG15]. The total space lower bound for random k-CNF formulas for $k \geqslant 4$, Theorem 7.2, was originally proved in [BGT14, BGT16]. The case $k = 3$ of those theorems was proved in [BBG⁺17].

The monomial space lower bound for *Tseitin formulas* over 4-regular random graphs was obtained by [FLM⁺13] as an application of [BG13, Theorem 1], which is a preliminary version of Theorem 4.2.

Theorem 7.8, the quadratic total space lower bound for Tseitin formulas, appeared in [Bon15, Bon16].

Part III
A Postlude

Chapter 8
Strong Size Lower Bounds for (a Subsystem of) Resolution

In this chapter we put the spotlight again on resolution size and in particular on its connection with conjectures about the exact complexity of the k-SAT problem, that is the conjectures known as the *Exponential Time Hypothesis* (ETH) and the *Strong Exponential Time Hypothesis* (SETH). We show a strong width lower bound (Theorem 8.1) and a strong size lower bound for a subsystem of resolution. The lower bounds are stronger than the one we could get immediately from the size-width inequality, eq. (2.10). This is made possible by a general hardness amplification result that relies on the combinatorial characterizations of size and width (Theorem 8.2).

8.1 SETH and Proof Complexity

We recall that the k-SAT problem is the problem of deciding whether a given k-CNF formula is satisfiable or not. There are several non-trivial algorithms known to solve k-SAT including the DPLL algorithms and the CDCL solvers we briefly saw in Sect. 1.1.4. Other examples of algorithms for k-SAT can be found for instance in [DGH+02, PPZ97, PPSZ05, Sch02]. Despite all of this however, the exact complexity of k-SAT under suitable hardness assumptions remains unknown. Formalizing what this complexity could be, [IP01] formulated the following two hypotheses: ETH and SETH.

The *Exponential Time Hypothesis* (ETH) states that the k-SAT problem requires exponential time, for every $k \geqslant 3$.

The *Strong Exponential Time Hypothesis* (SETH) states that the complexity of k-SAT grows as k increases and it approaches that of exhaustive search. More precisely let $\sigma_k = \inf\{\delta : k\text{-SAT can be solved in time } O(2^{\delta n}) \text{ for CNF formulas in } n \text{ variables}\}$. SETH states that $\lim_{k\to\infty} \sigma_k = 1$.

Both ETH and SETH are stronger than NP $\neq$ P and hence any proof of them is far beyond reach at the moment, but such hypotheses are important since they imply a plethora of fine-grained complexity results in the realm of *parameterized* complexity.

© Springer International Publishing AG, part of Springer Nature 2017
I. Bonacina, *Space in Weak Propositional Proof Systems*,
https://doi.org/10.1007/978-3-319-73453-8_8

We refer to [CFK+15] for more details on how these hypotheses are useful in such context.

Although a general proof of SETH and ETH seems out of reach at the moment one might ask whether these hypotheses hold for specific algorithms and classes of algorithms. That is whether there are k-CNF formulas on which the given algorithms run for at least $2^{(1-\varepsilon_k)n}$ steps in the case of SETH or $2^{\Omega(n)}$ in the case of ETH. Moreover, we can think about the run of a k-SAT algorithm on an unsatisfiable instance as a proof of its unsatisfiability; then, if the algorithm is structured enough, we can employ tools from proof complexity and obtain lower bounds on the running time.

For instance is well known that the run of a DPLL algorithm on an unsatisfiable k-CNF formula gives a tree-like resolution refutation of it (and vice versa). Therefore tree-like resolution size lower bounds imply DPLL running-time lower bounds. Similarly a resolution size lower bound will imply a lower bound on the running time of CDCL solvers, see Sect. 1.1.4. A resolution size lower bound for some unsatisfiable k-CNF formula F of the form $2^{\Omega(n)}$ will imply that no CDCL solver will ever be able to refute ETH. Similarly a resolution size lower bound of the form $2^{(1-\varepsilon_k)n}$ with $\varepsilon_k = o(1)$ implies that no CDCL solver will ever be able to disprove SETH.

Lower bounds of the form $2^{\Omega(n)}$ for k-CNF formulas in n variables have been known for a long time [Urq87]. If we restrict ourselves to tree-like resolution, [PI00] proved that there are k-CNF formulas F_n such that for every tree-like resolution refutation π,

$$S(\pi) \geqslant 2^{(1-\varepsilon_k)n} , \tag{8.1}$$

where $\varepsilon_k = O\left(k^{-\frac{1}{8}}\right)$. An analogous statement (for a different family of formulas and $\varepsilon_k = \tilde{O}\left(k^{-\frac{1}{3}}\right)$) was proved recently for regular resolution [BI13].

Here we prove an analogue of eq. (8.1) for a subsystem of resolution stronger than regular resolution that we call δ-regular resolution (Definition 8.1 and Corollary 8.2). Moreover we improve the asymptotic growth of ε_k in eq. (8.1) also in the case of tree-like resolution. Both results rely on the following strong width lower bound, which improves an analogous result in [BI13]. The connection between this result and a similar one in [BI13] is highlighted in the History section at the end of the chapter.

Theorem 8.1 ([BT16a]). *For any large n and k, there exist unsatisfiable k-CNF formulas F_n in n variables such that for every resolution refutation π of F_n*

$$W(\pi) \geqslant (1 - \zeta_k)n , \tag{8.2}$$

where $\zeta_k = \tilde{O}\left(k^{-\frac{1}{3}}\right)$.

The proof of this result is a bit long and technical and hence it is postponed to Sect. 8.3. Let's see first how to obtain the desired size lower bounds from this strong width lower bound, first for tree-like resolution and then for what we will call δ-regular resolution.

Corollary 8.1. *For any large enough k there exist unsatisfiable k-CNF formulas F_n in n variables such that for every* tree-like *resolution refutation π of F_n*

$$S(\pi) \geqslant 2^{(1-\varepsilon_k)n} , \tag{8.3}$$

where $\varepsilon_k = \widetilde{O}(k^{-\frac{1}{3}})$.

Proof. Let F_n be the unsatisfiable k-CNF formula in n variables coming from Theorem 8.1 with $w = (1 - \widetilde{O}\left(k^{-\frac{1}{3}}\right))n$. By eq. (2.17) then every tree-like resolution refutation π of F_n is such that

$$S(\pi) \geqslant 2^{w-k} , \tag{8.4}$$

hence the strong size lower bound follows. □

We can now consider the following generalization of regular resolution.

Definition 8.1 (δ-Regular Resolution). Given a CNF formula in n variables and $\delta \in [0,1]$, a resolution refutation π is δ-*regular* if there exists a witness function (see Definition 2.1) for π giving it a DAG structure where in each path the number of variables resolved multiple times is at most δn.

Clearly 0-regular resolution refutations are regular in the sense of Sect. 2.1 and 1-regular resolution refutations are just unconstrained resolution refutations.

To prove a strong size lower bound for δ-regular resolution we use a generalization of the 2-xorification technique we saw in Sect. 7.4. This is the concept of ℓ-xorification, a concept that already proved to be helpful in proof complexity, see for example [Nor13, Section 2.4] and [Ben02].

Definition 8.2 (ℓ-Xorification). Given a CNF formula F over a set of Boolean variables $X = \{x_1, \ldots, x_n\}$, the ℓ-*xorification* of F, $F[\oplus^{\ell}]$, is a CNF formula over a set of new Boolean variables $Y = \{y_i^j \, : \, i \in [n], j \in [\ell]\}$ that is obtained by replacing each occurrence of x_i in F with $y_i^1 \oplus \cdots \oplus y_i^{\ell}$ and then expanding the obtained propositional formula in CNF form.

Notice that if F is a k-CNF formula, then $F[\oplus^{\ell}]$ is a $k\ell$-CNF formula. An interesting property of xorifications is the following hardness amplification result, which can be used to give size lower bounds better than the ones we can get from the size-width inequality, see eq. (2.10).

Theorem 8.2 ([BT16b]). *Let F be an unsatisfiable CNF formula in n variables and let w, δ and ℓ be parameters. If every resolution refutation of F has width $> w$ then every δ-regular resolution refutation π' of $F[\oplus^{\ell}]$ is such that*

$$S(\pi') \geqslant 2^{(1-\varepsilon)w\ell} , \tag{8.5}$$

where $\varepsilon = \ell^{-1} \log\left(e^3 \ell n w^{-1}\right) + \delta n w^{-1} \log\left(e^3 \ell \delta^{-1}\right)$.

The proof of this result is a bit technical too and hence it is postponed to Sect. 8.2. From it and Theorem 8.1 there follows immediately a strong size lower bound for δ-regular resolution, for some small function δ.

Corollary 8.2. *For any large enough n and k there exists an unsatisfiable k-CNF formula F_n in n variables such that for every δ-regular resolution refutation π of F_n*

$$S(\pi) \geqslant 2^{(1-\varepsilon_k)n} , \tag{8.6}$$

where $\varepsilon_k = \delta = \widetilde{O}\left(k^{-\frac{1}{4}}\right)$.

Proof. Let $F'_{n'}$ be the k'-CNF formula in n' variables given by Theorem 8.1. In particular for every resolution refutation π of $F'_{n'}$

$$W(\pi) \geqslant (1 - \zeta_{k'})n' , \tag{8.7}$$

where $\zeta_{k'} = \widetilde{O}\left(1/\sqrt[3]{k'}\right)$. Then $F_n = F'_{n'}[\oplus^\ell]$ is a $k'\ell$-CNF formula on $n'\ell$ variables. Let $n = n'\ell$ and $k = k'\ell$. If we choose $\ell = \widetilde{\Theta}(\sqrt[3]{k'})$, $\delta = \widetilde{O}\left(1/\sqrt[3]{k'}\right)$ and $w = (1-\zeta_{k'})n'$ then, by Theorem 8.2, it follows that every δ-regular resolution refutation π of F_n has

$$S(\pi) \geqslant 2^{(1-\zeta_{k'})n'(\ell-\log(\frac{e^3 \ell n'}{w})-\frac{\delta \ell n'}{w}\log\frac{e^3 \ell}{\delta})} \tag{8.8}$$

$$= 2^{(1-\zeta_{k'})n\left(\ell-O(\log k')-\ell\widetilde{O}\left(1/\sqrt[3]{k'}\right)\right)} \tag{8.9}$$

$$= 2^{\left(1-\widetilde{O}\left(1/\sqrt[3]{k'}\right)\right)n'\ell} . \tag{8.10}$$

In particular eq. (8.9) follows from the choice of $\ell = \widetilde{\Theta}(\sqrt[3]{k'})$ and $\delta = \widetilde{O}\left(1/\sqrt[3]{k'}\right)$. To obtain the asymptotic behaviour of ε_k with respect to k, just observe that $k = k'\ell = \widetilde{\Theta}(k'^{\frac{4}{3}})$ and $\widetilde{O}\left(1/\sqrt[3]{k'}\right) = \widetilde{O}\left(k^{-\frac{1}{4}}\right)$, hence $\varepsilon_k = \widetilde{O}\left(k^{-\frac{1}{4}}\right)$. Similarly we get the asymptotic behaviour of δ as a function of k. $\square$

It then just remains to prove Theorem 8.1 and Theorem 8.2.

8.2 Game Characterizations of Width and Size

In this section we prove Theorem 8.2. But first we introduce a common terminology for two games: one characterizing resolution width [AD08], and one characterizing resolution size [Pud00].

Given an unsatisfiable CNF formula F and a set of Boolean assignments R containing the empty assignment, we define a game, $\mathfrak{G}(F, R)$, between two players Prover (he) and Delayer (she). At each step of the game $i \in \mathbb{N}$ the two players maintain a

Boolean assignment $\alpha_i \in R$, where $\alpha_0 \in R$ is the empty Boolean assignment. Then at step $i + 1$ the following moves take place:

1. Prover queries some variable $x \notin \text{dom}(\alpha_i)$.
2. Delayer answers $x = b$ for some bit $b \in \{0, 1\}$.
3. Prover sets $\alpha_{i+1} \in R$ such that $\alpha_{i+1} \subseteq \alpha_i \cup \{x = b\}$.

If at any point in the game α_i falsifies F then Prover wins; otherwise we say that Delayer wins. As usual, we say that Prover has a *winning strategy* for the game $\mathfrak{G}(F, R)$ if for any strategy of Delayer, he can play so that he wins the game. Otherwise we say that Delayer has a *winning strategy*. Notice that we can describe a strategy for Prover as a collection of pairs (α, x) where $\alpha \in R$ and x is a variable. The meaning of (α, x) is: "whenever in the game we reach the Boolean assignment α, query the variable x". A strategy for Delayer can be represented as a set of pairs of the form $((\alpha, x), b)$ with $\alpha \in R$, x a variable and $b \in \{0, 1\}$ with the meaning: "when given the Boolean assignment α, if Prover queries x then answer with $x = b$".

For a suitable choice of R the game $\mathfrak{G}(F, R)$ is exactly the one used by [AD08] to characterize the minimal width of resolution refutations of F. This is done by showing that the w-AD families introduced in Sect. 2.3.1 are winning strategies for Delayer for a suitable game $\mathfrak{G}(F, R)$ and then using the characterization of width from Theorem 2.5.

Theorem 8.3 ([AD08]). *Let F be an unsatisfiable CNF formula, w an integer and $\mathcal{W}$ the set of all possible Boolean assignments with a domain of size strictly less than w. Then* Delayer *has a winning strategy for $\mathfrak{G}(F, \mathcal{W})$ if and only if there exists a w-AD family for F. Due to this equivalence, we denote $\mathfrak{G}(F, \mathcal{W})$ as* width-$\mathfrak{G}(F, w)$.

Similarly, for size [Pud00] showed that one can characterize the minimal size of resolution refutations using this kind of game.

Theorem 8.4 ([Pud00]). *Let F be an unsatisfiable CNF formula. The following are equivalent:*

1. *there exists a set R of Boolean assignments such that $|R| \leqslant s$ and* Prover *has a winning strategy for $\mathfrak{G}(F, R)$;*
2. *there exists a resolution refutation π of F such that $\mathrm{S}(\pi) \leqslant s$.*

Essentially from a resolution refutation π of size S we can immediately construct a winning strategy for Prover for the game $\mathfrak{G}(F, R)$ with a set of Boolean assignments R of size S and vice versa: each play of the game $\mathfrak{G}(F, R)$ corresponds to a path in a DAG associated with π.

We want a similar result but for δ-regular resolution. To do so we need to limit the power of Prover.

Definition 8.3 ($\mathfrak{G}_\delta(F, R)$). Let F be a CNF formula in n variables, R a set of Boolean assignments and δ a parameter. If in each play of $\mathfrak{G}(F, R)$, Prover is allowed to re-query at most δn variables, we call the corresponding game $\mathfrak{G}_\delta(F, R)$.

We then have the following characterization of δ-regular resolution size.

Theorem 8.5 ([BT16b]). *Let F be an unsatisfiable CNF formula in n variables and* $\delta \in [0,1]$. *The following are equivalent*

1. *there exists a set R of Boolean assignments such that* $|R| \leqslant s$ *and* Prover *has a winning strategy for* $\mathfrak{G}_\delta(F,R)$;
2. *there exists a* δ-*regular resolution refutation* π *of F such that* $S(\pi) \leqslant s$.

In what follow we just use the implication $(2) \Rightarrow (1)$ and this is what we are going to sketch. The full proof can be found in [BT16b] although it is essentially the same as the proof of Theorem 8.4 from [Pud00], just adapted to δ-regular resolution.

Proof (sketch of (2) $\Rightarrow$ (1)). Given a refutation π of F of size at most s, for each clause C in π let α_C denote the Boolean assignment setting C to false on domain $\text{var}(C)$. We now define the set R to be $R = \{\alpha_C : C \in \pi\}$. By assumption we have that $|R| \leqslant s$. A winning strategy for Prover can be described simply taking a DAG associated with π and "reversing the direction of all edges." More precisely if a clause in π is the result of two clauses resolved over some variable x then just add to the strategy for Prover the pair (α_C, x).

Notice that each play of $\mathfrak{G}(F,R)$ corresponds to a path in π. Then, if π is δ-regular then in each play of the game the set of variables Prover is going to query many times is at most δn. So the strategy of Prover is for $\mathfrak{G}_\delta(F,R)$. □

We now have all the ingredients to prove the fact that the xorifications give rise to a hardness amplification, Theorem 8.2.

Restated Theorem 8.2 ([BT16b]) *Let F be an unsatisfiable CNF formula in n variables and let w, δ and ℓ be parameters. If every resolution refutation of F has width* $> w$ *then every δ-regular resolution refutation π' of $F[\oplus^\ell]$ is such that*

$$S(\pi') \geqslant 2^{(1-\varepsilon)w\ell}, \tag{8.5}$$

where $\varepsilon = \ell^{-1}\log\left(e^3 \ell n w^{-1}\right) + \delta n w^{-1} \log\left(e^3 \ell \delta^{-1}\right)$.

Proof. To prove the desired lower bound, by Theorem 8.5, it is enough to prove that whenever Prover wins $\mathfrak{G}_\delta(F[\oplus^\ell], R)$, the lower bound from eq. (8.5) is also a lower bound for $|R|$. By the assumption on the width needed to refute F we have, by Theorem 2.5, that there exists a w-AD family for F. Hence by Theorem 8.3, Delayer has a winning strategy σ for width-$\mathfrak{G}(F,w)$. We use this strategy to build many different strategies for Delayer when playing $\mathfrak{G}_\delta(F[\oplus^\ell], R)$. Then the idea is that now if Prover wants to win $\mathfrak{G}_\delta(F[\oplus^\ell], R)$, in particular, he must win against all the particular strategies we built and this will imply that $|R|$ must be large.

Let's start by fixing some notations. Let $X = \{x_1, \ldots, x_n\}$ be the variables of F and let $Y = \{y_1^1, \ldots, y_1^\ell, \ldots, y_n^1, \ldots, y_n^\ell\}$ be the variables of $F[\oplus^\ell]$ where by construction we have that semantically $x_i \equiv \bigoplus_{j=1}^\ell y_i^j$. To avoid confusion we call the x_i variables, *x-variables*, and the y_i^j variables, *y-variables*. Moreover we say that the y-variables $y_i^1, \ldots, y_i^\ell$ form a *block* of variables corresponding to the x-variable x_i.

With each Boolean assignment α over Y there is naturally associated a Boolean assignment α' over X:

$$\alpha'(x_i) = \begin{cases} \alpha(y_i^1) \oplus \cdots \oplus \alpha_r(y_i^\ell) & \text{if } \forall j \in [\ell],\ y_i^j \in \text{dom}(\alpha), \\ \star & \text{otherwise}. \end{cases} \qquad (8.11)$$

Now, given a winning strategy σ for Delayer in the game width-$\mathfrak{G}(F, w)$ and any total Boolean assignment β over the y-variables we can build a strategy σ_β for Delayer in the game $\mathfrak{G}_\delta(F[\oplus^\ell], R)$ intuitively answering according to β, except when this would be a "stupid" idea; then use σ. Formally, given a Boolean assignment α over Y and a variable $y_i^j \notin \text{dom}(\alpha)$ we want to define a $b \in \{0,1\}$ that Delayer should answer, that is we want to find a reasonable $((\alpha, y_i^j), b)$ to put in σ_β:

1. if $((\alpha', x_i), 0) \notin \sigma$ and $((\alpha', x_i), 1) \notin \sigma$ then put $((\alpha, y_i^j), 0) \in \sigma_\beta$ (or $((\alpha, y_i^j), 1) \in \sigma_\beta$, it doesn't really matter);
2. if $((\alpha', x_i), b) \in \sigma$ and there exists $j' \neq j$ such that $y_i^{j'} \notin \text{dom}(\alpha)$ then put $((\alpha, y_i^j), \beta(y_i^j)) \in \sigma_\beta$;
3. otherwise if $((\alpha', x_i), b') \in \sigma$ and for each $j' \neq j$, $y_i^{j'} \in \text{dom}(\alpha)$ let $b \in \{0,1\}$ such that

$$b \oplus \bigoplus_{j' \neq j} \alpha(y_i^j) = b'. \qquad (8.12)$$

Then put $((\alpha, y_i^j), b) \in \sigma_\beta$.

It is easy to see that for each total Boolean assignment β over Y, σ_β is a winning strategy for Delayer in the game width-$\mathfrak{G}(F[\oplus^\ell], w\ell)$. Since we are assuming that Prover wins against all of those strategies this means that for each total Boolean assignment β over Y there exists some Boolean assignment $\alpha_\beta \in R$ such that $\text{dom}(\alpha_\beta) \geqslant w\ell$ and moreover at least w blocks of y-variables are completely inside $\text{dom}(\alpha_\beta)$. Without loss of generality we assume that each α_β has domain consisting of exactly w blocks of y-variables. That is we (possibly) simply ignore some of the variables in $\text{dom}(\alpha_\beta)$ and only consider w blocks inside it completely set.

In the remaining part of the proof we show that there must be 'many' distinct such Boolean assignments α_β. Let $B \in \binom{[n]}{w}$ and let

$$S_B = \left\{ \beta \in \{0,1\}^Y \ : \ \forall i \in B\ x_i \in \text{dom}(\alpha'_\beta) \right\}, \qquad (8.13)$$

that is, in other words, S_B is the set of total Boolean assignments β over Y such that the corresponding α_β sets exactly all the y-variables y_i^j with $i \in B$ and $j \in [\ell]$. Now, clearly, for any possible $B \in \binom{[n]}{w}$

$$|R| \geqslant \left| \left\{ \alpha_\beta \ : \ \beta \in \{0,1\}^Y \right\} \right| \geqslant \left| \left\{ \alpha_\beta \ : \ \beta \in S_B \right\} \right|. \qquad (8.14)$$

Let $A = \{\alpha_\beta \, : \, \beta \in S_B\}$. Let's show that there exists some B^* such that eq. (8.14) for B^* gives the desired lower bound of eq. (8.5).

There are $2^{n\ell}$ possible total Boolean assignments β over Y and $\binom{n}{w}$ possible sets $B \in \binom{[n]}{w}$, hence, by the pigeonhole principle, there exists a set $B^* \in \binom{[n]}{w}$ such that

$$|S_{B^*}| \geqslant \frac{2^{n\ell}}{\binom{n}{w}} \, . \tag{8.15}$$

Let S'_{B^*} be the set of all the restrictions of Boolean assignments in S_{B^*} to the set $\{y_i^j \, : \, i \in B^* \text{ and } j \in [\ell]\}$. Then

$$|S_{B^*}| \leqslant |S'_{B^*}| \cdot 2^{n\ell - \ell |B^*|} = |S'_{B^*}| \cdot 2^{n\ell - w\ell} \, , \tag{8.16}$$

and, by eq. (8.15), then

$$|S'_{B^*}| \geqslant \frac{2^{w\ell}}{\binom{n}{w}} \, . \tag{8.17}$$

We now have that both S'_{B^*} and A consist of Boolean assignments with domain $\{y_i^j \, : \, i \in B^* \text{ and } j \in [\ell]\}$. We show that A cannot be too small compared to S'_{B^*}. Intuitively, this will be due to the fact that the βs we start with are very different and the fact that Prover in each play of the game cannot re-query too many variables.

Now a Boolean assignment $\beta' \in S'_{B^*}$ may not be in A, for instance due to some y-variable that was set last, and hence σ_β prescribes to not answer according to β' but preserving σ. This could lead indeed to the fact that some different strategies σ_{β_0} and σ_{β_1} lead to the same Boolean assignment in A, that is $\alpha_{\beta_0} = \alpha_{\beta_1}$. More generally if in the i-th block Z_i variables are queried multiple times then β and α_β may differ in $|Z_i| + 1$ variables in the i-block. In each play of the game Prover can, by hypothesis, re-query at most $\delta n\ell$ variables, hence, by counting the variables where a Boolean assignment in A and a Boolean assignment in S'_{B^*} might differ, we have that

$$|S'_{B^*}| \leqslant |A| \cdot \sum_{Z \in \binom{Y}{\delta n\ell}} \prod_{i \in B^*} 2^{|Z_i| + 1} \binom{\ell}{|Z_i| + 1} \, , \tag{8.18}$$

where $Z_i = Z \cap \{y_i^1, \ldots, y_i^\ell\}$. Finally we can simplify this last expression as follows:

$$|S'_{B^*}| \overset{(8.18)}{\leqslant} |A| \cdot \sum_{Z \in \binom{Y}{\delta n \ell}} \prod_{i \in B^*} 2^{|Z_i|+1} \binom{\ell}{|Z_i|+1} \qquad (8.19)$$

$$\leqslant |A| \cdot \sum_{Z \in \binom{Y}{\delta n \ell}} \prod_{i \in B^*} \left(\frac{e^2 \ell}{|Z_i|+1} \right)^{|Z_i|+1} \qquad (8.20)$$

$$\leqslant |A| \cdot \sum_{Z \in \binom{Y}{\delta n \ell}} \left(\frac{\sum_{i \in B^*} e^2 \ell}{\sum_{i \in B^*}(|Z_i|+1)} \right)^{\sum_{i \in B^*}(|Z_i|+1)} \qquad (8.21)$$

$$\leqslant |A| \cdot \binom{\ell n}{\delta \ell n} \cdot \left(e^2 \ell \right)^{\delta \ell n + w}. \qquad (8.22)$$

The inequality (8.21) follows from the weighted AM-GM inequality[1] and the inequality (8.22) follows from the fact that $w \leqslant \sum_{i \in B^*}(|Z_i|+1) \leqslant \delta \ell n + w$ and the hypothesis that $|B^*| = w$. Then, putting everything together, we have that

$$|R| \overset{(8.14)}{\geqslant} |A| \geqslant \frac{|S'_{B^*}|}{\binom{n \ell}{\delta \ell n} (e^2 \ell)^{\delta \ell n + w}} \overset{(8.17)}{\geqslant} \frac{2^{w \ell}}{\binom{n}{w} \binom{\ell n}{\delta \ell n} (e^2 \ell)^{\delta \ell n + w}} \qquad (8.24)$$

$$\geqslant \frac{2^{w \ell}}{\left(\frac{en}{w} \right)^w \left(\frac{e}{\delta} \right)^{\delta \ell n} (e^2 \ell)^{\delta \ell n + w}} \qquad (8.25)$$

$$= 2^{w \left(\ell - \log\left(\frac{e^3 \ell n}{w} \right) - \frac{\delta \ell n}{w} \log\left(\frac{e^3 \ell}{\delta} \right) \right)}. \qquad \square \qquad (8.26)$$

8.3 A Strong Width Lower Bound

The last missing step now is to prove Theorem 8.1, that is we want to construct unsatisfiable formulas whose require very large resolution width refutations. Such a construction is in [BI13] and improved in [BT16a]. Here we show the later improved version.

Let p be a prime, $\mathbb{F}_p$ be the finite field with p elements and $\mathbf{v} = (v_1, v_2, \ldots)$ be a vector over $\mathbb{F}_p$. Then by $\operatorname{supp}(\mathbf{v})$ we denote the indices of $\mathbf{v}$ with non-zero entries mod p, that is $\operatorname{supp}(\mathbf{v}) = \{i : v_i \not\equiv 0 \mod p\}$. We use the letter E, with subscripts, to denote linear equations mod p, that is expressions of the form $\sum_j a_j z_j \equiv b \mod p$ with $a_j, b \in \mathbb{F}_p$. The set of indices j having non-zero a_js is $\operatorname{supp}(E)$.

---

[1] The *weighted Arithmetic-Geometric Mean inequality* says that given non-negative numbers $a_1, \ldots, a_n$ and non-negative weights $w_1, \ldots, w_n$ then

$$\prod_i a_i^{w_i} \leqslant \left(\frac{\sum_i w_i a_i}{w} \right)^w, \qquad (8.23)$$

where $w = \sum_i w_i$. We applied this inequality with $a_i = \frac{e^2 \ell}{|Z_i|+1}$ and $w_i = |Z_i|+1$.

Given two linear equalities E and E', their sum $E + E'$ is what one should expect, that is if E is $\sum_j a_j z_j \equiv b \mod p$ and E' is $\sum_j a'_j z_j \equiv b' \mod p$ then $E + E'$ is $\sum_j (a_j + a'_j) z_j \equiv (b + b') \mod p$. Similarly we define rE for $r \in \mathbb{F}$ and $\sum_i r_i E_i$ for a set of linear equations $\{E_i\}_i$ and $r_i \in \mathbb{F}$.

In [BI13] it is proven that there are unsatisfiable systems of linear equations mod p with good expansion properties.

Proposition 8.1 ([BI13, Lemma 4.2]). *Let p be a sufficiently large prime. There exists a set $\mathcal{E} := \{E_1, \ldots, E_{n+1}\}$ consisting of linear equations in n variables over $\mathbb{F}_p$ and there exist $\delta = O(1/p)$ and $\theta = \widetilde{O}(1/p)$ such that*

1. for each $E_i \in \mathcal{E}$, $|\text{supp}(E_i)| \leqslant p^2$;
2. $\mathcal{E}$ is unsatisfiable but no subset of at most $3\delta n$ equations from $\mathcal{E}$ is unsatisfiable;
*3. (**Expansion**) for every $\mathbf{v} = (v_1, \ldots, v_m) \in \mathbb{F}_p^{n+1}$, if $|\text{supp}(\mathbf{v})| \in [\delta n, 3\delta n]$ then*

$$\left| \text{supp} \left(\sum_{i=1}^{n+1} v_i E_i \right) \right| \geqslant (1 - \theta)n . \qquad \Box \tag{8.27}$$

Now the idea is to encode the system of mod p equations $\mathcal{E}$ coming from the previous proposition using Boolean variables in a redundant way.

Lemma 8.1. *Let p be a sufficiently large prime, $\theta = \widetilde{O}(1/p)$ and let $u = \theta^{-1} \log^2 p$. There exists a function $g : \{0,1\}^u \to \{0,1\}^{\log p}$ such that for any Boolean assignment α with $|\text{dom}(\alpha)| \leqslant u - \log^2 p$ we have that $g\!\restriction_\alpha$ is surjective.*

Proof. Let g be a random function that assigns to every $x \in \{0,1\}^u$ a value in $\{0,1\}^{\log p}$ independently and uniformly at random. We bound the probability that there exist a $y \in \{0,1\}^{\log p}$ and a Boolean assignment α with $|\text{dom}(\alpha)| = u - \log^2 p$ such that $y \notin \text{Img}(g\!\restriction_\alpha)$. Let A be this event. A bound on $\Pr[A]$ is easily given as follows:

$$\Pr[A] \leqslant 2^{\log p} \binom{u}{\log^2 p} 2^{u - \log^2 p} \left(1 - \frac{1}{p}\right)^{2^{\log^2 p}} \tag{8.28}$$

$$\leqslant p\theta^{-\log^2 p} 2^{u - \log^2 p} e^{\log^2 p - \frac{1}{p} 2^{\log^2 p}} \tag{8.29}$$

$$= o(1) , \tag{8.30}$$

since $\theta = \widetilde{O}(1/p)$. The inequality in (8.28) follows by the union bound, since once we fixed $y \in \{0,1\}^{\log p}$ and a Boolean assignment α such that $|\text{dom}(\alpha)| = u - \log^2 p$ then

$$\Pr[y \notin \text{Img}(g\!\restriction_\alpha)] \leqslant (1 - 1/p)^{2^{\log^2 p}} . \tag{8.31}$$

Then there must exist at least one function g realizing the complementary event that we bounded. This function works also for each α such that $|\text{dom}(\alpha)| \leqslant u - \log^2 p$. $\Box$

Given the function $g : \{0,1\}^{\theta^{-1}\log^2 p} \to \{0,1\}^{\log p}$ provided by the previous proposition we can then redundantly encode systems of linear equations as follows.

Let $Z = \{z_1,\ldots,z_n\}$ be a set of variables taking values over $\mathbb{F}_p$. We encode the mod p value of each variable z_i using $u = \theta^{-1}\log^2 p$ new Boolean variables $X = \{x_{i1},\ldots,x_{iu}\}$:

$$z_i = \sum_{j=1}^{\log p} 2^{j-1} g_j(x_{i1},\ldots,x_{iu}),\qquad (8.32)$$

where g_j represents the projection of g on the j-th coordinate. Hence a mod p linear equation E in n variables, say

$$\sum_i a_i z_i \equiv b \mod p,\qquad (8.33)$$

can be transformed into a Boolean function E^b using eq. (8.32) and $nu = n\theta^{-1}\log^2 p$ Boolean variables x_{ij}:

$$\sum_{j=1}^n a_{ij} \sum_{k=1}^{\log p} 2^{k-1} g_k(x_{i1},\ldots,x_{iu}) \equiv b_i \mod p.\qquad (8.34)$$

Moreover if $|\mathrm{supp}(a_1,\ldots,a_n)| \leqslant d$ then the natural Boolean encoding of this function as a CNF formula turns out to be a (du)-CNF formula.

Let's proceed then to prove Theorem 8.1. This proof is essentially an adaptation of an analogous proof from [BI13] to this context, see the History section for more details.

Restated Theorem 8.1 ([BT16a]) *For any large n and k, there exist unsatisfiable k-CNF formulas F_n in n variables such that for every resolution refutation π of F_n*

$$W(\pi) \geqslant (1 - \zeta_k)n,\qquad (8.2)$$

where $\zeta_k = \widetilde{O}\left(k^{-\frac{1}{3}}\right)$.

Proof. Let p be a sufficiently large prime and let $\mathcal{E} := \{E_1,\ldots,E_m\}$, $\delta = O(1/p)$ and $\theta = \widetilde{O}(1/p)$ be the set of linear equations in n variables over $\mathbb{F}_p$ and the parameters from Proposition 8.1. Let $u = \theta^{-1}\log^2 p$ and $g : \{0,1\}^u \to \{0,1\}^{\log p}$ be the function from Lemma 8.1. The CNF formula F we consider is the natural encoding of the Boolean function, that is

$$\bigwedge_{i=1}^m E_i^b\qquad (8.35)$$

as a CNF formula, where the E_i^b are the Boolean encodings of the $E_i \in \mathcal{E}$ as in eq. (8.34).

Since for each i we have $|\mathrm{supp}(E_i)| \leqslant p^2$, then F is a $(p^2\theta^{-1}\log^2 p)$-CNF formula in $N = nu = n\theta^{-1}\log^2 p$ variables. We prove that for each resolution refutation π of F it holds that

$$W(\pi) \geqslant (1 - 2\theta)N \ . \tag{8.36}$$

Recalling that $\theta = \tilde{O}(1/p)$ this implies immediately eq. (8.2).

   To prove eq. (8.36) we use a "medium complexity clause" type of argument. Let $\mathcal{E}^b := \{E_i^b \ : \ E_i \in \mathcal{E}\}$ and for each clause C let $\mu(C)$ be the following complexity measure:

$$\mu(C) = \min\left\{|S| \ : \ S \subseteq \mathcal{E}^b \text{ and } S \vDash C\right\} \ . \tag{8.37}$$

A clause C has medium complexity (with respect to μ) if $\mu(C) \in \left(\frac{3}{2}\delta n, 3\delta n\right]$. We have that given any clauses C_1, C_2 and C_3 in π such that $C_1 \wedge C_2 \vDash C_3$ then $\mu(c_3) \leqslant \mu(C_1) + \mu(C_2)$ and there will be at least one clause of medium complexity in π. Let C be one such clause. We show that $|C| \geqslant (1 - 2\theta)N$, hence proving eq. (8.36).

   Suppose, for sake of contradiction, that $|C| < (1 - 2\theta)N$. Let the variables in the set $Z = \{z_1, \ldots, z_n\}$ be Z-*variables* and, similarly, the variables in the set $X = \{x_{ij} \ : \ i \in [n] \text{ and } j \in [u]\}$ be X-*variables*. Let α be the Boolean assignment over the X-variables setting C to false with domain exactly $\text{var}(C)$. That is in particular $|\text{dom}(\alpha)| < (1 - 2\theta)N$. We say that a Z-variable z_i is *free* if

$$|\text{dom}(\alpha) \cap \{x_{i1}, \ldots, x_{iu}\}| \leqslant u - \log^2 p \ . \tag{8.38}$$

Let ξ be the number of Z-variables that are free. First we prove that $\xi > \theta n$. We have both an upper and a lower bound for $N - |\text{dom}(\alpha)|$:

$$2\theta N < N - |\text{dom}(\alpha)| \leqslant (n - \xi)(u - (u - \log^2 p)) + u\xi \ . \tag{8.39}$$

Hence

$$2\theta N < n \log^2 p - \xi \log^2 p + u\xi \ . \tag{8.40}$$

Now if $\xi \leqslant \theta n$ a contradiction follows immediately recalling that $N = un$ and $\theta N = n \log^2 p$.

   We say that a Boolean assignment $\sigma : X \to \{0, 1, \star\}$ is a completion of α if it extends α and has domain $\{x_{ij} \ : \ z_i \text{ is not free}\}$. Let A be the set of all Boolean assignments over X that are completions of α. Recalling the definition of the X-variables in terms of the Z-variables in eq. (8.32), we have that each $\sigma \in A$ naturally defines a Boolean assignment $\sigma' : Z \to \mathbb{F}_p \cup \{\star\}$ with domain $\{z_i \ : \ z_i \text{ is not free}\}$. More precisely

$$\sigma'(z_i) = \begin{cases} \sum_{j=1}^{\log p} 2^{j-1} g_j(\sigma(x_{i1}), \ldots, \sigma(x_{iu})) & \text{if } z_i \text{ is not free} \ , \\ \star & \text{otherwise} \ . \end{cases} \tag{8.41}$$

So, for each $\sigma \in A$, the Z-variables that are free are exactly, by construction, the ones not in the domain of σ' and for each $\sigma \in A$, σ sets C to false. As observed we have that the number of free variables $\xi > c\theta n$ and hence

$$|\text{dom}(\sigma')| < n - \theta n = (1 - \theta)n \ . \tag{8.42}$$

As C is of medium complexity with respect to μ, there exists some set of equations $S \subseteq \mathcal{E}^b$ such that $S \vDash C$, $|S| \in (\frac{3}{2}\delta n, 3\delta n]$ and S is minimal with respect to inclusion. Let $S' = \{E \; : \; E^b \in S\}$. This implies that for each possible $\sigma \in A$ of the form described above, both $S\restriction_\sigma$ and $S'\restriction_{\sigma'}$ are unsatisfiable. Moreover, by the minimality of S, for each equation $E \in S'$ there exists some $\sigma \in A$ such that $E\restriction_{\sigma'}$ is not a trivial constraint, that is a constraint that is always satisfied. The fact that, for each $\sigma \in A$, $S'\restriction_{\sigma'}$ is unsatisfiable means exactly that for all $\sigma \in A$ there exists some $\mathbf{v} = (v_1, \ldots, v_{n+1}) \in \mathbb{F}_p^{n+1}$ (dependent on σ) with $|\mathrm{supp}(\mathbf{v})| \leqslant |S| = |S'|$ and such that $\sum_{i=1}^{n+1} v_i E_i\restriction_{\sigma'}$ is unsatisfiable. Hence for each $\sigma \in A$,

$$\mathrm{supp}(\sum_{i=1}^{n+1} v_i E_i) \subseteq \mathrm{dom}(\sigma') \,, \tag{8.43}$$

otherwise we could use the variables not fixed by σ' to satisfy the equality $\sum_i v_i E_i\restriction_{\sigma'}$. Moreover by what we observed before, for each $E \in S$ there exists some $\sigma \in A$ such that $E\restriction_{\sigma'}$ does not trivialize and hence E_i will appear in the sum above for that σ.

Given $\sigma \in A$, let $E^\sigma = \sum_i v_i E_i$, where $\mathbf{v} = (v_1, \ldots, v_m)$ depends on σ as in the sum above. Then we take a random linear combination of all the E^σs for all the possible $\sigma \in A$: let $\sum_{\sigma \in A} \alpha_\sigma E^\sigma$ be this combination. Again we have that

$$\mathrm{supp}(\sum_{\sigma \in A} \alpha_\sigma E^\sigma) \subseteq \bigcup_{\sigma \in A} \mathrm{dom}(\sigma') \,. \tag{8.44}$$

Each $E_i \in S'$ appears in the previous sum since, as already observed, for each E_i there exists some $\sigma \in A$ such that E_i appears in E^σ. Moreover, the coefficient of each $E_i \in S$ is uniformly random, and hence by averaging, there exists a linear combination such that at least $(1 - 1/p)\frac{3}{2}\delta n \geqslant \delta n$ of the E_i have non-zero coefficient. This contradicts the expansion property of $\mathcal{E}$ as we have that

$$|\mathrm{supp}(\sum_{\sigma \in A} \alpha_\sigma E^\sigma)| \leqslant |\bigcup_{\sigma \in A} \mathrm{dom}(\sigma')| < (1 - \theta)n \,, \tag{8.45}$$

where the last inequality follows from the inequality in (8.42) and the fact that all the $\sigma \in A$ have the same domain. □

8.4 Open Problems

The obvious challenge here is to prove a strong size lower bound for resolution or any proof system stronger than δ-regular resolution where we already have some exponential-size lower bound. In particular there is the following natural question.

Question 8.1. Is there any unsatisfiable k-CNF formula F_n in n variables such that for large enough n and k we have that for every resolution refutation π of F_n

$$S(\pi) \geqslant 2^{(1-\varepsilon_k)n} \,, \tag{8.46}$$

where $\varepsilon_k \to 0$ as $k \to \infty$?

For (tree-like) resolution we might be asking whether there are formulas for which the upper bound of Theorem 2.1 is asymptotically tight, that is more formally the following.

Question 8.2. Is there any unsatisfiable k-CNF formula F_n in n variables such that for every *tree-like* resolution refutation π of F_n

$$\log_2 S(\pi) \geqslant \left(1 - O\left(k^{-1}\right)\right) n \,? \tag{8.47}$$

Another question that arises quite naturally is whether δ-regular resolution is exponentially stronger than regular resolution, or more precisely, what is the growth rate of δ as a function of k needed to guarantee this property? Is $\frac{1}{2}$-regular resolution p-equivalent to resolution?

History

The results shown in this chapter are mainly based on [BT16b, BT16a]; in particular Theorem 8.1 and Theorem 8.2 are based on analogous results from those papers.

The proof of Theorem 8.1 has a lot of analogies with a similar result from [BI13]. The main difference between our theorem and the one from [BI13] is a different way of encoding the linear equations mod p from Proposition 8.1. In [BI13] this is done using a sum of roughly p^2 Boolean variables. The key property of this representation is the following: let $z = \sum_{i=0}^{p^2} x_i$, where the x_i are Boolean variables, then even setting a lot of variables (that is $p^2 - p$) we still can obtain all possible $\mathbb{F}_p$ values for z by setting the remaining variables.

In other words what is really needed in [BI13] is a function that can extract $\log p$ bits even after many bits in the input are fixed. The way we address this is to just show that a random function satisfies this property, see Lemma 8.1. We then use this function instead of the sum of p^2 Boolean variables. Then the same argument from [BI13] goes through.

References

[ABLM08] Carlos Ansótegui, Maria Luisa Bonet, Jordi Levy, and Felip Manyà. Measuring the hardness of SAT instances. In Dieter Fox and Carla P. Gomes, editors, *Proceedings of the Twenty-Third AAAI Conference on Artificial Intelligence, AAAI 2008, Chicago, Illinois, USA, July 13-17, 2008*, pages 222–228. AAAI Press, 2008.

[ABRW02] Michael Alekhnovich, Eli Ben-Sasson, Alexander A. Razborov, and Avi Wigderson. Space complexity in propositional calculus. *SIAM J. Comput.*, 31(4):1184–1211, 2002.

[AD08] Albert Atserias and Víctor Dalmau. A combinatorial characterization of resolution width. *J. Comput. Syst. Sci.*, 74(3):323–334, 2008.

[AFT11] Albert Atserias, Johannes Klaus Fichte, and Marc Thurley. Clause-learning algorithms with many restarts and bounded-width resolution. *J. Artif. Intell. Res. (JAIR)*, 40:353–373, 2011.

[AJPU07] Michael Alekhnovich, Jan Johannsen, Toniann Pitassi, and Alasdair Urquhart. An exponential separation between regular and general resolution. *Theory of Computing*, 3(1):81–102, 2007.

[Ajt94] Miklós Ajtai. The complexity of the pigeonhole principle. *Combinatorica*, 14(4):417–433, 1994.

[Ale11] Michael Alekhnovich. Lower bounds for k-DNF resolution on random 3-CNFs. *Computational Complexity*, 20(4):597–614, 2011.

[ALN14] Albert Atserias, Massimo Lauria, and Jakob Nordström. Narrow proofs may be maximally long. In *IEEE 29th Conference on Computational Complexity, CCC 2014, Vancouver, BC, Canada, June 11-13, 2014*, pages 286–297. IEEE, 2014.

[AR01] Michael Alekhnovich and Alexander A. Razborov. Lower bounds for polynomial calculus: Non-binomial case. In *42nd Annual Symposium on Foundations of Computer Science, FOCS 2001, 14-17 October 2001, Las Vegas, Nevada, USA*, pages 190–199. IEEE Computer Society, 2001.

[Ats04] Albert Atserias. On sufficient conditions for unsatisfiability of random formulas. *J. ACM*, 51(2):281–311, 2004.

[BBG+17] Patrick Bennett, Ilario Bonacina, Nicola Galesi, Tony Huynh, Mike Molloy, and Paul Wollan. Space proof complexity for random 3-CNFs. *Information and Computation*, 255:165–176, 2017.

[BBI12] Paul Beame, Christopher Beck, and Russell Impagliazzo. Time-space tradeoffs in resolution: superpolynomial lower bounds for superlinear space. In Howard J. Karloff and Toniann Pitassi, editors, *Proceedings of the 44th Symposium on Theory of Computing Conference, STOC 2012, New York, NY, USA, May 19 - 22, 2012*, pages 213–232. ACM, 2012.

[BD09] Michael Brickenstein and Alexander Dreyer. PolyBoRi: A framework for Gröbner-basis computations with Boolean polynomials. *Journal of Symbolic Computation*, 44(9):1326–1345, 2009.

© Springer International Publishing AG, part of Springer Nature 2017   119
I. Bonacina, *Space in Weak Propositional Proof Systems*,
https://doi.org/10.1007/978-3-319-73453-8

[BDG+09]   M. Brickenstein, A. Dreyer, G. Greuel, M. Wedler, and O. Wienand. New developments in the theory of Gröbner bases and applications to formal verification. *Journal of Pure and Applied Algebra*, 213(8):1612–1635, 2009.

[Bea94]   Paul Beame. A Switching Lemma Primer. Technical report, Department of Computer Science and Engineering, University of Washington, 1994. UW-CSE-95-07-01.

[Ben02]   Eli Ben-Sasson. Size space tradeoffs for resolution. In John H. Reif, editor, *Proceedings on 34th Annual ACM Symposium on Theory of Computing, May 19-21, 2002, Montréal, Québec, Canada*, pages 457–464. ACM, 2002.

[BFU93]   Andrei Z. Broder, Alan M. Frieze, and Eli Upfal. On the satisfiability and maximum satisfiability of random 3-CNF formulas. In Vijaya Ramachandran, editor, *Proceedings of the Fourth Annual ACM/SIGACT-SIAM Symposium on Discrete Algorithms, 25-27 January 1993, Austin, Texas*, pages 322–330. ACM/SIAM, 1993.

[BG99]   Maria Luisa Bonet and Nicola Galesi. A study of proof search algorithms for resolution and polynomial calculus. In *40th Annual Symposium on Foundations of Computer Science, FOCS '99, 17-18 October, 1999, New York, NY, USA*, pages 422–432. IEEE Computer Society, 1999.

[BG01]   Maria Luisa Bonet and Nicola Galesi. Optimality of size-width tradeoffs for resolution. *Computational Complexity*, 10(4):261–276, 2001.

[BG03]   Eli Ben-Sasson and Nicola Galesi. Space complexity of random formulae in resolution. *Random Struct. Algorithms*, 23(1):92–109, 2003.

[BG13]   Ilario Bonacina and Nicola Galesi. Pseudo-partitions, transversality and locality: A Combinatorial Characterization for the Space Measure in Algebraic Proof Systems. In *4th Conf. Innov. Theor. Comput. Sci. – ITCS*, pages 455–472, 2013.

[BG15]   Ilario Bonacina and Nicola Galesi. A Framework for Space Complexity in Algebraic Proof Systems. *J. ACM*, 62(3):1–20, June 2015.

[BGIP99] Samuel R. Buss, Dima Grigoriev, Russell Impagliazzo, and Toniann Pitassi. Linear gaps between degrees for the polynomial calculus modulo distinct primes. In Jeffrey Scott Vitter, Lawrence L. Larmore, and Frank Thomson Leighton, editors, *Proceedings of the Thirty-First Annual ACM Symposium on Theory of Computing, May 1-4, 1999, Atlanta, Georgia, USA*, pages 547–556. ACM, 1999.

[BGL10]   Olaf Beyersdorff, Nicola Galesi, and Massimo Lauria. A lower bound for the pigeonhole principle in tree-like resolution by asymmetric prover-delayer games. *Inf. Process. Lett.*, 110(23):1074–1077, 2010.

[BGL13]   Olaf Beyersdorff, Nicola Galesi, and Massimo Lauria. A characterization of tree-like resolution size. *Inf. Process. Lett.*, 113(18):666–671, 2013.

[BGT14]   Ilario Bonacina, Nicola Galesi, and Neil Thapen. Total Space in Resolution. In *55th Annu. Symp. Found. Comput. Sci. – FOCS*, volume 38, pages 641–650, oct 2014.

[BGT16]   Ilario Bonacina, Nicola Galesi, and Neil Thapen. Total Space in Resolution. *SIAM J. Comput.*, 45(5):1894–1909, Jan 2016.

[BI10]   Eli Ben-Sasson and Russell Impagliazzo. Random CNF's are hard for the polynomial calculus. *Computational Complexity*, 19(4):501–519, 2010.

[BI13]   Christopher Beck and Russell Impagliazzo. Strong ETH holds for regular resolution. In Dan Boneh, Tim Roughgarden, and Joan Feigenbaum, editors, *Symposium on Theory of Computing Conference, STOC'13, Palo Alto, CA, USA, June 1-4, 2013*, pages 487–494. ACM, 2013.

[BIK+92] Paul Beame, Russell Impagliazzo, Jan Krajíček, Toniann Pitassi, Pavel Pudlák, and Alan R. Woods. Exponential lower bounds for the pigeonhole principle. In S. Rao Kosaraju, Mike Fellows, Avi Wigderson, and John A. Ellis, editors, *Proceedings of the 24th Annual ACM Symposium on Theory of Computing, May 4-6, 1992, Victoria, British Columbia, Canada*, pages 200–220. ACM, 1992.

[BIK+94] Paul Beame, Russell Impagliazzo, Jan Krajíček, Toniann Pitassi, and Pavel Pudlák. Lower bound on Hilbert's Nullstellensatz and propositional proofs. In *35th Annual Symposium on Foundations of Computer Science, Santa Fe, New Mexico, USA, 20-22 November 1994*, pages 794–806. IEEE Computer Society, 1994.

[BIK+97]   Samuel R. Buss, Russell Impagliazzo, Jan Krajíček, Pavel Pudlák, Alexander A. Razborov, and Jiri Sgall. Proof complexity in algebraic systems and bounded depth Frege systems with modular counting. *Computational Complexity*, 6(3):256–298, 1997.

[BJS97]   Roberto J. Bayardo Jr. and Robert Schrag. Using CSP look-back techniques to solve real-world SAT instances. In Benjamin Kuipers and Bonnie L. Webber, editors, *Proceedings of the Fourteenth National Conference on Artificial Intelligence and Ninth Innovative Applications of Artificial Intelligence Conference, AAAI 97, IAAI 97, July 27-31, 1997, Providence, Rhode Island*, pages 203–208. AAAI Press / The MIT Press, 1997.

[BK14]   Olaf Beyersdorff and Oliver Kullmann. Unified characterisations of resolution hardness measures. In Carsten Sinz and Uwe Egly, editors, *Theory and Applications of Satisfiability Testing - SAT 2014 - 17th International Conference, Held as Part of the Vienna Summer of Logic, VSL 2014, Vienna, Austria, July 14-17, 2014. Proceedings*, volume 8561 of *Lecture Notes in Computer Science*, pages 170–187. Springer, 2014.

[BKPS98]   Paul Beame, Richard M. Karp, Toniann Pitassi, and Michael E. Saks. On the complexity of unsatisfiability proofs for random k-CNF formulas. In Jeffrey Scott Vitter, editor, *Proceedings of the Thirtieth Annual ACM Symposium on the Theory of Computing, Dallas, Texas, USA, May 23-26, 1998*, pages 561–571. ACM, 1998.

[BKPS02]   Paul Beame, Richard M. Karp, Toniann Pitassi, and Michael E. Saks. The efficiency of resolution and Davis–Putnam procedures. *SIAM J. Comput.*, 31(4):1048–1075, 2002.

[Bla37]   Archie Blake. *Canonical Expressions in Boolean Algebra*. PhD thesis, 1937. University of Chicago.

[BN08]   Eli Ben-Sasson and Jakob Nordström. Understanding space in resolution: optimal lower bounds and exponential trade-offs. In Peter Bro Miltersen, Rüdiger Reischuk, Georg Schnitger, and Dieter van Melkebeek, editors, *Computational Complexity of Discrete Problems, 14.09. - 19.09.2008*, volume 08381 of *Dagstuhl Seminar Proceedings*. Schloss Dagstuhl - Leibniz-Zentrum für Informatik, Germany, 2008.

[BN09]   Eli Ben-Sasson and Jakob Nordström. A space hierarchy for k-DNF resolution. *Electronic Colloquium on Computational Complexity (ECCC)*, 16:47, 2009.

[BN11]   Eli Ben-Sasson and Jakob Nordström. Understanding space in proof complexity: Separations and trade-offs via substitutions. In Bernard Chazelle, editor, *Innovations in Computer Science - ICS 2010, Tsinghua University, Beijing, China, January 7-9, 2011. Proceedings*, pages 401–416. Tsinghua University Press, 2011.

[BN16]   Christoph Berkholz and Jakob Nordström. Supercritical space-width trade-offs for resolution. In Ioannis Chatzigiannakis, Michael Mitzenmacher, Yuval Rabani, and Davide Sangiorgi, editors, *43rd International Colloquium on Automata, Languages, and Programming, ICALP 2016, July 11-15, 2016, Rome, Italy*, volume 55 of *LIPIcs*, pages 57:1–57:14. Schloss Dagstuhl - Leibniz-Zentrum fuer Informatik, 2016.

[BNT13]   Chris Beck, Jakob Nordström, and Bangsheng Tang. Some trade-off results for polynomial calculus: extended abstract. In Dan Boneh, Tim Roughgarden, and Joan Feigenbaum, editors, *Symposium on Theory of Computing Conference, STOC'13, Palo Alto, CA, USA, June 1-4, 2013*, pages 813–822. ACM, 2013.

[Bon15]   Ilario Bonacina. *Space in weak propositional proof systems*. PhD thesis, Sapienza University of Rome, 2015.

[Bon16]   Ilario Bonacina. Total Space in Resolution is at Least Width Squared. In *43rd International Colloquium on Automata, Languages, and Programming – ICALP*, volume 55, pages 56:1–56:13, 2016.

[BP96]   Paul Beame and Toniann Pitassi. Simplified and improved resolution lower bounds. In *37th Annual Symposium on Foundations of Computer Science, FOCS '96, Burlington, Vermont, USA, 14-16 October, 1996*, pages 274–282. IEEE Computer Society, 1996.

[BP97]   Samuel R. Buss and Toniann Pitassi. Resolution and the weak pigeonhole principle. In Mogens Nielsen and Wolfgang Thomas, editors, *Computer Science Logic, 11th International Workshop, CSL '97, Annual Conference of the EACSL, Aarhus, Denmark, August 23-29, 1997, Selected Papers*, volume 1414 of *Lecture Notes in Computer Science*, pages 149–156. Springer, 1997.

[BP01]   Paul Beame and Toniann Pitassi. Propositional proof complexity: Past, present, and
            future. In *Current Trends in Theoretical Computer Science*, pages 42–70. 2001.

[BS01]   Eli Ben-Sasson. *Expansion in Proof Complexity*. PhD thesis, 2001. Hebrew University.

[BS14]   Boaz Barak and David Steurer. Sum-of-squares proofs and the quest toward optimal
            algorithms. *Proceedings of International Congress of Mathematicians (ICM)*, IV:509–
            533, 2014.

[BT15]   Ilario Bonacina and Navid Talebanfard. Strong ETH and Resolution via Games and
            the Multiplicity of Strategies. In *10th International Symposium on Parameterized and
            Exact Computation – IPEC*, pages 248–257, 2015.

[BT16a]   Ilario Bonacina and Navid Talebanfard. Improving resolution width lower bounds for
            k-CNFs with applications to the Strong Exponential Time Hypothesis. *Inf. Process.
            Lett.*, 116(2):120–124, 2016.

[BT16b]   Ilario Bonacina and Navid Talebanfard. Strong ETH and Resolution via Games and the
            Multiplicity of Strategies. *Algorithmica*, pages 1–13, October 2016.

[Bus87]   Samuel R. Buss. Polynomial size proofs of the propositional pigeonhole principle. *J.
            Symb. Log.*, 52(4):916–927, 1987.

[BW01]   Eli Ben-Sasson and Avi Wigderson. Short proofs are narrow - resolution made simple.
            J. ACM, 48(2):149–169, 2001.

[CCT87]   William Cook, Collette R. Coullard, and György Turán. On the complexity of cutting-
            plane proofs. *Discrete Applied Mathematics*, 18(1):25–38, 1987.

[CEI96]   Matthew Clegg, Jeff Edmonds, and Russell Impagliazzo. Using the Groebner basis
            algorithm to find proofs of unsatisfiability. In Gary L. Miller, editor, *Proceedings of
            the Twenty-Eighth Annual ACM Symposium on the Theory of Computing, Philadelphia,
            Pennsylvania, USA, May 22-24, 1996*, pages 174–183. ACM, 1996.

[CF90]   Ming-Te Chao and John V. Franco. Probabilistic analysis of a generalization of the unit-
            clause literal selection heuristics for the k satisfiability problem. *Inf. Sci.*, 51(3):289–314,
            1990.

[CFK+15]   Marek Cygan, Fedor V. Fomin, Lukasz Kowalik, Daniel Lokshtanov, Dániel Marx,
            Marcin Pilipczuk, Michal Pilipczuk, and Saket Saurabh. *Parameterized Algorithms*.
            Springer, 2015.

[Chv73]   Vašek Chvátal. Edmonds polytopes and a hierarchy of combinatorial problems. *Discrete
            Mathematics*, 4(4):305 – 337, 1973.

[CKT91]   Peter Cheeseman, Bob Kanefsky, and William M. Taylor. Where the really hard
            problems are. In John Mylopoulos and Raymond Reiter, editors, *Proceedings of the
            12th International Joint Conference on Artificial Intelligence. Sydney, Australia, August
            24-30, 1991*, pages 331–340. Morgan Kaufmann, 1991.

[CLO97]   David A. Cox, John Little, and Donal O'Shea. *Ideals, varieties, and algorithms - an
            introduction to computational algebraic geometry and commutative algebra (2. ed.)*.
            Undergraduate texts in mathematics. Springer, 1997.

[CR79]   Stephen A. Cook and Robert A. Reckhow. The relative efficiency of propositional proof
            systems. *J. Symb. Log.*, 44(1):36–50, 1979.

[CR92]   Vasek Chvátal and Bruce A. Reed. Mick gets some (the odds are on his side). In *33rd
            Annual Symposium on Foundations of Computer Science, Pittsburgh, Pennsylvania,
            USA, 24-27 October 1992*, pages 620–627. IEEE Computer Society, 1992.

[CS88]   Vasek Chvátal and Endre Szemerédi. Many hard examples for resolution. *J. ACM*,
            35(4):759–768, 1988.

[DGH+02]   Evgeny Dantsin, Andreas Goerdt, Edward A. Hirsch, Ravi Kannan, Jon M. Kleinberg,
            Christos H. Papadimitriou, Prabhakar Raghavan, and Uwe Schöning. A deterministic
            $(2 - 2/(k + 1))^n$ algorithm for k-SAT based on local search. *Theor. Comput. Sci.*,
            289(1):69–83, 2002.

[DGH+05]   Heidi E. Dixon, Matthew L. Ginsberg, David K. Hofer, Eugene M. Luks, and Andrew J.
            Parkes. Generalizing Boolean satisfiability III: implementation. *J. Artif. Intell. Res.
            (JAIR)*, 23:441–531, 2005.

[DGLP04] Heidi E. Dixon, Matthew L. Ginsberg, Eugene M. Luks, and Andrew J. Parkes. Generalizing Boolean satisfiability II: theory. *J. Artif. Intell. Res. (JAIR)*, 22:481–534, 2004.

[DGP04]   Heidi E. Dixon, Matthew L. Ginsberg, and Andrew J. Parkes. Generalizing Boolean satisfiability I: background and survey of existing work. *J. Artif. Intell. Res. (JAIR)*, 21:193–243, 2004.

[DLL62]   Martin Davis, George Logemann, and Donald W. Loveland. A machine program for theorem-proving. *Commun. ACM*, 5(7):394–397, 1962.

[dlV01]   Wenceslas Fernandez de la Vega. Random 2-SAT: results and problems. *Theor. Comput. Sci.*, 265(1-2):131–146, 2001.

[DP60]   Martin Davis and Hilary Putnam. A computing procedure for quantification theory. *J. ACM*, 7(3):201–215, 1960.

[DSS15]   Jian Ding, Allan Sly, and Nike Sun. Proof of the satisfiability conjecture for large k. In Rocco A. Servedio and Ronitt Rubinfeld, editors, *Proceedings of the Forty-Seventh Annual ACM on Symposium on Theory of Computing, STOC 2015, Portland, OR, USA, June 14-17, 2015*, pages 59–68. ACM, 2015.

[EGM04]   Juan Luis Esteban, Nicola Galesi, and Jochen Messner. On the complexity of resolution with bounded conjunctions. *Theor. Comput. Sci.*, 321(2-3):347–370, 2004.

[ET01]   Juan Luis Esteban and Jacobo Torán. Space bounds for resolution. *Inf. Comput.*, 171(1):84–97, 2001.

[FLM+13] Yuval Filmus, Massimo Lauria, Mladen Mikša, Jakob Nordström, and Marc Vinyals. Towards an understanding of polynomial calculus: New separations and lower bounds - (extended abstract). In Fedor V. Fomin, Rusins Freivalds, Marta Z. Kwiatkowska, and David Peleg, editors, *Automata, Languages, and Programming - 40th International Colloquium, ICALP 2013, Riga, Latvia, July 8-12, 2013, Proceedings, Part I*, volume 7965 of *Lecture Notes in Computer Science*, pages 437–448. Springer, 2013.

[FLN+15] Yuval Filmus, Massimo Lauria, Jakob Nordström, Noga Ron-Zewi, and Neil Thapen. Space complexity in polynomial calculus. *SIAM J. Comput.*, 44(4):1119–1153, 2015.

[FPPR17] Noah Fleming, Denis Pankratov, Toniann Pitassi, and Robert Robere. Random CNFs are Hard for Cutting Planes. *Electronic Colloquium on Computational Complexity (ECCC)*, 24:45, 2017.

[Fri98]   Ehud Friedgut. Sharp thresholds of graph properties, and the k-SAT problem. *J. Amer. Math. Soc*, 12:1017–1054, 1998.

[FS96]   Alan M. Frieze and Stephen Suen. Analysis of two simple heuristics on a random instance of k-SAT. *J. Algorithms*, 20(2):312–355, 1996.

[GH01]   Dima Grigoriev and Edward A. Hirsch. Algebraic proof systems over formulas. *Electronic Colloquium on Computational Complexity (ECCC)*, 8(11), 2001.

[GHP01]   Dima Grigoriev, Edward A. Hirsch, and Dmitrii V. Pasechnik. Complexity of semi-algebraic proofs. *Electronic Colloquium on Computational Complexity (ECCC)*, (103), 2001.

[GL10a]   Nicola Galesi and Massimo Lauria. On the automatizability of polynomial calculus. *Theory Comput. Syst.*, 47(2):491–506, 2010.

[GL10b]   Nicola Galesi and Massimo Lauria. Optimality of size-degree tradeoffs for polynomial calculus. *ACM Trans. Comput. Log.*, 12(1):4, 2010.

[Goe96]   Andreas Goerdt. A threshold for unsatisfiability. *J. Comput. Syst. Sci.*, 53(3):469–486, 1996.

[Gom63]   Ralph E. Gomory. An algorithm for integer solutions to linear programs. *Recent Advances in Mathematical Programming*, pages 269–302, 1963.

[GP14]   Joshua A. Grochow and Toniann Pitassi. Circuit complexity, proof complexity, and polynomial identity testing. In *55th IEEE Annual Symposium on Foundations of Computer Science, FOCS 2014, Philadelphia, PA, USA, October 18-21, 2014*, pages 110–119. IEEE Computer Society, 2014.

[GPT15]   Nicola Galesi, Pavel Pudlák, and Neil Thapen. The space complexity of cutting planes refutations. In David Zuckerman, editor, *30th Conference on Computational Complexity,*

CCC 2015, June 17-19, 2015, Portland, Oregon, USA, volume 33 of *LIPIcs*, pages 433–447. Schloss Dagstuhl - Leibniz-Zentrum fuer Informatik, 2015.

[Gri01]   Dima Grigoriev. Linear lower bound on degrees of Positivstellensatz calculus proofs for the parity. *Theor. Comput. Sci.*, 259(1-2):613–622, 2001.

[Hak85]   Armin Haken. The intractability of resolution. *Theor. Comput. Sci.*, 39:297–308, 1985.

[Hås87]   Johan Håstad. *Computational Limitations of Small-depth Circuits*. MIT Press, Cambridge, MA, USA, 1987.

[HC99]   Armin Haken and Stephen A. Cook. An exponential lower bound for the size of monotone real circuits. *J. Comput. Syst. Sci.*, 58(2):326–335, 1999.

[HLW06]   Shlomo Hoory, Nathan Linial, and Avi Wigderson. Expander graphs and their applications. *Bull. Amer. Math. Soc.*, 43(4):439–561, 2006.

[HP17]   Pavel Hrubes and Pavel Pudlák. Random formulas, monotone circuits, and interpolation. *Electronic Colloquium on Computational Complexity (ECCC)*, 24:42, 2017.

[HY87]   Wenqi Huang and Xiangdong Yu. A DNF without Regular Shortest Consensus Path. *SIAM Journal on Computing*, 16(5):836–840, 1987.

[IP01]   Russell Impagliazzo and Ramamohan Paturi. On the complexity of k-SAT. *J. Comput. Syst. Sci.*, 62(2):367–375, 2001.

[IPS99]   Russell Impagliazzo, Pavel Pudlák, and Jiri Sgall. Lower bounds for the polynomial calculus and the Gröbner basis algorithm. *Computational Complexity*, 8(2):127–144, 1999.

[Jeř05]   Emil Jeřábek. *Weak pigeonhole principle, and randomized computation*. PhD thesis, Faculty of Mathematics and Physics, Charles University, Prague, 2005.

[JMNZ12]   Matti Järvisalo, Arie Matsliah, Jakob Nordström, and Stanislav Zivny. Relating proof complexity measures and practical hardness of SAT. In Michela Milano, editor, *Principles and Practice of Constraint Programming - 18th International Conference, CP 2012, Québec City, QC, Canada, October 8-12, 2012. Proceedings*, volume 7514 of *Lecture Notes in Computer Science*, pages 316–331. Springer, 2012.

[Juk12]   Stasys Jukna. *Boolean Function Complexity - Advances and Frontiers*, volume 27 of *Algorithms and combinatorics*. Springer, 2012.

[KKKS98]   Lefteris M. Kirousis, Evangelos Kranakis, Danny Krizanc, and Yannis C. Stamatiou. Approximating the unsatisfiability threshold of random formulas. *Random Struct. Algorithms*, 12(3):253–269, 1998.

[KL94]   Hans Kleine Büning and Theodor Lettmann. *Aussagenlogik - Deduktion und Algorithmen*. Leitfäden und Monographien der Informatik. Teubner, 1994.

[Koz77]   Dexter Kozen. Lower bounds for natural proof systems. In *18th Annual Symposium on Foundations of Computer Science, Providence, Rhode Island, USA, 31 October - 1 November 1977*, pages 254–266. IEEE Computer Society, 1977.

[KPW95a]   Jan Krajíček, Pavel Pudlák, and Alan R. Woods. An exponential lower bound to the size of bounded depth Frege proofs of the pigeonhole principle. *Random Struct. Algorithms*, 7(1):15–40, 1995.

[KPW95b]   Jan Krajíček, Pavel Pudlák, and Alan Woods. Exponential lower bounds to the size of bounded depth Frege proofs of the pigeonhole principle. *Random Structures and Algorithms*, 7(1):15–39, 1995.

[Kra94]   Jan Krajícek. Lower bounds to the size of constant-depth propositional proofs. *J. Symb. Log.*, 59(1):73–86, 1994.

[Kra95]   Jan Krajíček. *Bounded Arithmetic, Propositional Logic and Complexity Theory*. Cambridge University Press, 1995.

[Kra97]   Jan Krajíček. Interpolation theorems, lower bounds for proof systems, and independence results for bounded arithmetic. *J. Symb. Log.*, 62(2):457–486, 1997.

[Kra01]   Jan Krajíček. On the weak pigeonhole principle. *Fund. Math.*, 170(1-2):123–140, 2001. Dedicated to the memory of Jerzy Łoś.

[Kra09]   Jan Krajíček. Propositional proof complexity I. *Online manuscript*, 2009. http://www.karlin.mff.cuni.cz/~krajicek/ds1.ps.

[Kul99]   Oliver Kullmann. Investigating a general hierarchy of polynomially decidable classes of CNF's based on short tree-like resolution proofs. *Electronic Colloquium on Computational Complexity (ECCC)*, (41), 1999.

[Kul00]   Oliver Kullmann. An improved version of width restricted resolution. In *AMAI*, 2000.

[Kul04]   Oliver Kullmann. Upper and lower bounds on the complexity of generalised resolution and generalised constraint satisfaction problems. *Ann. Math. Artif. Intell.*, 40(3-4):303–352, 2004.

[MMZ$^+$01] Matthew W. Moskewicz, Conor F. Madigan, Ying Zhao, Lintao Zhang, and Sharad Malik. Chaff: Engineering an efficient SAT solver. In *Proceedings of the 38th Design Automation Conference, DAC 2001, Las Vegas, NV, USA, June 18-22, 2001*, pages 530–535. ACM, 2001.

[MN15]   Mladen Mikša and Jakob Nordström. A generalized method for proving polynomial calculus degree lower bounds. In David Zuckerman, editor, *30th Conference on Computational Complexity, CCC 2015, June 17-19, 2015, Portland, Oregon, USA*, volume 33 of *LIPIcs*, pages 467–487. Schloss Dagstuhl - Leibniz-Zentrum fuer Informatik, 2015.

[MRW05]   Peter Bro Miltersen, Jaikumar Radhakrishnan, and Ingo Wegener. On converting CNF to k-DNF. *Theor. Comput. Sci.*, 347(1-2):325–335, 2005.

[NH13]   Jakob Nordström and Johan Håstad. Towards an optimal separation of space and length in resolution. *Theory of Computing*, 9:471–557, 2013.

[Nor09]   Jakob Nordström. Narrow proofs may be spacious: Separating space and width in resolution. *SIAM J. Comput.*, 39(1):59–121, 2009.

[Nor13]   Jakob Nordström. Pebble games, proof complexity, and time-space trade-offs. *Logical Methods in Computer Science*, 9(3), 2013.

[Nor15]   Jakob Nordström. On the interplay between proof complexity and SAT solving. *ACM SIGLOG News*, 2(3):19–44, August 2015.

[PBI93]   Toniann Pitassi, Paul Beame, and Russell Impagliazzo. Exponential lower bounds for the pigeonhole principle. *Computational Complexity*, 3:97–140, 1993.

[PD11]   Knot Pipatsrisawat and Adnan Darwiche. On the power of clause-learning SAT solvers as resolution engines. *Artif. Intell.*, 175(2):512–525, 2011.

[PI00]   Pavel Pudlák and Russell Impagliazzo. A lower bound for DLL algorithms for k-SAT (preliminary version). In David B. Shmoys, editor, *Proceedings of the Eleventh Annual ACM-SIAM Symposium on Discrete Algorithms, January 9-11, 2000, San Francisco, CA, USA.*, pages 128–136. ACM/SIAM, 2000.

[Pit96]   Toniann Pitassi. Algebraic propositional proof systems. In Neil Immerman and Phokion G. Kolaitis, editors, *Descriptive Complexity and Finite Models, Proceedings of a DIMACS Workshop, January 14-17, 1996, Princeton University*, volume 31 of *DIMACS Series in Discrete Mathematics and Theoretical Computer Science*, pages 215–244. American Mathematical Society, 1996.

[Pit11]   Toniann Pitassi. Propositional proof complexity: A survey on the state of the art, including some recent results. In *Proceedings of the 26th Annual IEEE Symposium on Logic in Computer Science, LICS 2011, June 21-24, 2011, Toronto, Ontario, Canada*, page 119. IEEE Computer Society, 2011.

[PPSZ05]   Ramamohan Paturi, Pavel Pudlák, Michael E. Saks, and Francis Zane. An improved exponential-time algorithm for k-SAT. *J. ACM*, 52(3):337–364, 2005.

[PPZ97]   Ramamohan Paturi, Pavel Pudlák, and Francis Zane. Satisfiability coding lemma. In *38th Annual Symposium on Foundations of Computer Science, FOCS*, pages 566–574, 1997.

[PU95]   Toniann Pitassi and Alasdair Urquhart. The complexity of the Hajoś calculus. *SIAM J. Discrete Math.*, 8(3):464–483, 1995.

[Pud97]   Pavel Pudlák. Lower bounds for resolution and cutting plane proofs and monotone computations. *J. Symb. Log.*, 62(3):981–998, 1997.

[Pud00]   Pavel Pudlák. Proofs as games. *The American Mathematical Monthly*, 107(6):541–550, 2000.

[Pud08]   Pavel Pudlák. Twelve problems in proof complexity. In Edward A. Hirsch, Alexander A.
           Razborov, Alexei L. Semenov, and Anatol Slissenko, editors, *Computer Science -
           Theory and Applications, Third International Computer Science Symposium in Russia,
           CSR 2008, Moscow, Russia, June 7-12, 2008, Proceedings*, volume 5010 of *Lecture
           Notes in Computer Science*, pages 13–27. Springer, 2008.

[Raz98]   Alexander A. Razborov. Lower bounds for the polynomial calculus. *Computational
           Complexity*, 7(4):291–324, 1998.

[Raz01]   Alexander A. Razborov. Proof complexity of pigeonhole principles. In Werner Kuich,
           Grzegorz Rozenberg, and Arto Salomaa, editors, *Developments in Language Theory,
           5th International Conference, DLT 2001, Vienna, Austria, July 16-21, 2001, Revised
           Papers*, volume 2295 of *Lecture Notes in Computer Science*, pages 100–116. Springer,
           2001.

[Raz03]   Alexander A. Razborov. Resolution lower bounds for the weak functional pigeonhole
           principle. *Theor. Comput. Sci.*, 1(303):233–243, 2003.

[Raz04]   Ran Raz. Resolution lower bounds for the weak pigeonhole principle. *J. ACM*,
           51(2):115–138, 2004.

[Raz16a]   Alexander Razborov. A new kind of tradeoffs in propositional proof complexity. *J.
           ACM*, 63(2):16:1–16:14, April 2016.

[Raz16b]   Alexander A. Razborov. On space and depth in resolution. *Electronic Colloquium on
           Computational Complexity (ECCC)*, 23:184, 2016.

[Rec75]   Robert A. Reckhow. *On the lengths of proofs in the propositional calculus*. PhD thesis,
           1975.

[RM09]   Olivier Roussel and Vasco M. Manquinho. Pseudo-boolean and cardinality constraints.
           In Armin Biere, Marijn Heule, Hans van Maaren, and Toby Walsh, editors, *Handbook
           of Satisfiability*, volume 185 of *Frontiers in Artificial Intelligence and Applications*,
           pages 695–733. IOS Press, 2009.

[Rob65]   John Alan Robinson. A machine-oriented logic based on the resolution principle. *J.
           ACM*, 12(1):23–41, 1965.

[Rob16]   Alexander Roberts. Tree matchings. *arXiv preprint arXiv: 1612.01694*, 12 2016.

[RT08]   Ran Raz and Iddo Tzameret. The strength of multilinear proofs. *Computational
           Complexity*, 17(3):407–457, 2008.

[Sch97]   Uwe Schöning. Resolution proofs, exponential bounds, and Kolmogorov complexity.
           In Igor Prívara and Peter Ruzicka, editors, *Mathematical Foundations of Computer
           Science 1997, 22nd International Symposium, MFCS'97, Bratislava, Slovakia, August
           25-29, 1997, Proceedings*, volume 1295 of *Lecture Notes in Computer Science*, pages
           110–116. Springer, 1997.

[Sch02]   Uwe Schöning. A probabilistic algorithm for k-SAT based on limited local search and
           restart. *Algorithmica*, 32(4):615–623, 2002.

[Seg07]   Nathan Segerlind. The complexity of propositional proofs. *Bulletin of Symbolic Logic*,
           13(4):417–481, 2007.

[Sho67]   Joseph R. Shoenfield. *Mathematical Logic*. Addison-Wesley, Reading, Massachusetts,
           1967.

[SS99]   João P. Marques Silva and Karem A. Sakallah. GRASP: A search algorithm for
           propositional satisfiability. *IEEE Trans. Computers*, 48(5):506–521, 1999.

[Stå96]   Gunnar Stålmarck. Short resolution proofs for a sequence of tricky formulas. *Acta
           Informatica*, 33(3):277–280, 1996.

[Ste73]   Gilbert Stengle. A Nullstellensatz and Positivstellensatz in semialgebraic geometry.
           math. Ann., 207:87–97, 1973.

[Tha14]   Neil Thapen. A trade-off between length and width in resolution. *Electronic Colloquium
           on Computational Complexity (ECCC)*, 21:137, 2014.

[Tse83]   G.S. Tseitin. On the complexity of derivation in propositional calculus. In Jörg H.
           Siekmann and Graham Wrightson, editors, *Automation of Reasoning*, Symbolic Com-
           putation, pages 466–483. Springer Berlin Heidelberg, 1983.

[Urq87]   Alasdair Urquhart. Hard examples for resolution. *J. ACM*, 34(1):209–219, 1987.

[Urq95]   Alasdair Urquhart. The complexity of propositional proofs. *Bulletin of Symbolic Logic*, 1(4):425–467, 1995.
[Urq11a]   Alasdair Urquhart. The depth of resolution proofs. *Studia Logica*, 99(1-3):349–364, 2011.
[Urq11b]   Alasdair Urquhart. A near-optimal separation of regular and general resolution. *SIAM J. Comput.*, 40(1):107–121, 2011.
[Zit02]   Michele Zito. An upper bound on the space complexity of random formulae in resolution. *ITA*, 36(4):329–339, 2002.

Index

© Springer International Publishing AG, part of Springer Nature 2017
I. Bonacina, *Space in Weak Propositional Proof Systems*,
https://doi.org/10.1007/978-3-319-73453-8

Printed in the United States
By Bookmasters